I0019905

Khaled A. Harras

Challenged Networks

Khaled A. Harras

Challenged Networks

Protocol and Architectural Challenges in Delay and Disruption Tolerant Networks

VDM Verlag Dr. Müller

Impressum/Imprint (nur für Deutschland/ only for Germany)
Bibliografische Information der Deutschen Nationalbibliothek: Die Deutsche Nationalbibliothek
verzeichnet diese Publikation in der Deutschen Nationalbibliografie; detaillierte bibliografische
Daten sind im Internet über http://dnb.d-nb.de abrufbar.
Alle in diesem Buch genannten Marken und Produktnamen unterliegen warenzeichen-, marken-
oder patentrechtlichem Schutz bzw. sind Warenzeichen oder eingetragene Warenzeichen der
jeweiligen Inhaber. Die Wiedergabe von Marken, Produktnamen, Gebrauchsnamen,
Handelsnamen, Warenbezeichnungen u.s.w. in diesem Werk berechtigt auch ohne besondere
Kennzeichnung nicht zu der Annahme, dass solche Namen im Sinne der Warenzeichen- und
Markenschutzgesetzgebung als frei zu betrachten wären und daher von jedermann benutzt
werden dürften.

Coverbild: www.purestockx.com

Verlag: VDM Verlag Dr. Müller Aktiengesellschaft & Co. KG
Dudweiler Landstr. 125 a, 66123 Saarbrücken, Deutschland
Telefon +49 681 9100-698, Telefax +49 681 9100-988, Email: info@vdm-verlag.de
Zugl.: Santa Barbara, University of California, 2007

Herstellung in Deutschland:
Schaltungsdienst Lange o.H.G., Zehrensdorfer Str. 11, D-12277 Berlin
Books on Demand GmbH, Gutenbergring 53, D-22848 Norderstedt
Reha GmbH, Dudweiler Landstr. 99, D- 66123 Saarbrücken
ISBN: 978-3-639-08377-4

Imprint (only for USA, GB)
Bibliographic information published by the Deutsche Nationalbibliothek: The Deutsche
Nationalbibliothek lists this publication in the Deutsche Nationalbibliografie; detailed
bibliographic data are available in the Internet at http://dnb.d-nb.de.
Any brand names and product names mentioned in this book are subject to trademark, brand or
patent protection and are trademarks or registered trademarks of their respective holders. The use
of brand names, product names, common names, trade names, product descriptions etc. even
without
a particular marking in this works is in no way to be construed to mean that such names may be
regarded as unrestricted in respect of trademark and brand protection legislation and could thus
be used by anyone.

Cover image: www.purestockx.com

Publisher:
VDM Verlag Dr. Müller Aktiengesellschaft & Co. KG
Dudweiler Landstr. 125 a, 66123 Saarbrücken, Germany
Phone +49 681 9100-698, Fax +49 681 9100-988, Email: info@vdm-verlag.de

Copyright © 2008 VDM Verlag Dr. Müller Aktiengesellschaft & Co. KG and licensors
All rights reserved. Saarbrücken 2008

Produced in USA and UK by:
Lightning Source Inc., 1246 Heil Quaker Blvd., La Vergne, TN 37086, USA
Lightning Source UK Ltd., Chapter House, Pitfield, Kiln Farm, Milton Keynes, MK11 3LW, GB
BookSurge, 7290 B. Investment Drive, North Charleston, SC 29418, USA
ISBN: 978-3-639-08377-4

To my father's soul and my mother's spirit.

To those that fell, picked themselves up, and

rose above all obstacles in order to succeed.

Acknowledgements

To begin with, I owe a lot of my gratitude and appreciation to my advisor, Kevin Almeroth, for the tremendous amount of support and guidance that he has provided me with during my journey as a graduate student. His vision, patience, and ability to lead by example, have all been inspirational to me. His constant push for me to excel in various aspects in parallel, and to constantly raise the bar, helped me further realize my true potential. In short, through five years of closely interacting with Kevin, he has helped me acquire an invaluable set of skills, principles, and values, that I believe will continue to be precious assets I will carry throughout my life.

I would also like to thank the other members of my committee, Elizabeth Belding, and Chandra Krintz. Their guidance, feedback, and support have been greatly helpful with my research.

I also thank my friends and fellow graduate students at the Networking and Multimedia Systems Lab (NMSL) and the Mobility Management and Networking (MOMENT) lab. Their general support and wonderful sense of humor were very soothing at times when things were rough. Special thanks also goes to Mike Wittie and Caitlin Holman for their collaboration in a couple of research projects.

I am grateful for all those friends that supported me, stood by me and shared the happy and sad times of this long journey. A very warm gratitude also goes to those friends that were proud of me and supported me despite the huge distance in place and time between us. Also, my fellow friends and teammates in the Computer Science department and Ultimate Frisbee team deserve a special recognition. The closeness and solidarity developed between us over the past five years has been very helpful in lifting my spirit at many times when I was down.

There is no doubt that I would have never made it here in the first place if it weren't for my parents' love, guidance, dedication, and support. In the end, no words can describe the amount of love, appreciation, and gratitude that I carry in my heart for a very special person that stood by me in many times, suffered a lot with me, and provided me with endless unconditional love. This person will always have a special place my heart.

Curriculum Vitæ

Khaled A. Harras

Education

2007	M.S./Ph.D. in Computer Science, University of California, Santa Barbara.
2001	B.S. in Computer Science, The American University in Cairo, Egypt.

Experience

2002 – 2007	Graduate Research Assistant, University of California, Santa Barbara.
2005, 2006	Software Engineer Summer Intern, Citrix Systems.
2004	Instructor, University of California, Santa Barbara.
2003 – 2004	Teaching Assistant, University of California, Santa Barbara.
2002	Teaching Associate, The American University in Cairo.
2001	Research Assistant, The American University in Cairo.
1999	LAN Administrator Assistant, UNESCO, Egypt.
1998	Web Design Assistant, UNESCO, Egypt.
1997	Support Engineer Assistant, Giza Systems Engineering, Egypt.

Services

2006 – 2007	Web Chair, ACM Workshop on Challenged Networks (CHANTS).
2002 – 2007	Student member of ACM and IEEE.
1999 – 2001	Contest Committe Head, then President, ACM Student Chapter, The American University in Cairo.

Awards and Honors

2003 – 2004	Outstanding Educator award, University of California, Santa Barbara.
2003 – 2004	Three time recipient of the Outstanding TA award, University of California, Santa Barbara.
2002 – 2003	Graduate Fellowship, University of California, Santa Barbara.
2001 – 2002	President's Cup award; Ahmed H. Zewail prize; and Exemplary Student award; all from The American University in Cairo.

| 1997 – 2001 | Four Academic Achievement Scholarships; Three Presidential Merit Scholarships; Three Computer Science Honor Certificates; ACM Certificate of Achievement; all from The American University in Cairo. |

Selected Publications

- Mike P. Wittie, Khaled A. Harras, Kevin C. Almeroth, and Elizabeth M. Belding, "Cloud Routing: Self-Aware Traffic Routing in ParaNets-Enabled Delay Tolerant Mobile Networks". (To submit)

- Khaled A. Harras, Mike P. Wittie, Kevin C. Almeroth, and Elizabeth M. Belding, "The Vision and Challenges of A Parallel Networks Architecture (ParaNets) for Delay and Disruption Tolerant Networks". (To submit).

- Khaled A. Harras and Kevin C. Almeroth, "The Taxonomy of Challenged Networks". (Survey to submit).

- Khaled A. Harras and Kevin C. Almeroth, "Messenger Scheduling in Disconnected Clustered Mobile Networks". (Submitted to Journal).

- Khaled A. Harras and Kevin C. Almeroth, "Controlled Flooding in Disconnected Sparse Mobile Networks". Accepted to the *Journal on Wireless Communication and Mobile Computing (WCMC)*.

- Khaled A. Harras, Mike P. Wittie, Kevin C. Almeroth and Elizabeth M. Belding, "ParaNets: A Parallel Network Architecture for Challenged Networks", *IEEE Workshop on Mobile Computing Systems and Applications (Hotmobile)*, Tucson, AZ, February 2000.

- Caitlin Holman, Khaled A. Harras, Kevin C. Almeroth and Anderson Lam, "A Proactive Data Bundling System for Intermittent Mobile Connection", *IEEE International Conference on Sensor and Ad Hoc Communications and Networks (SECON)*, Reston, VA, September 2006.

- Khaled A. Harras and Kevin C. Almeroth, "Inter-Regional Messenger Scheduling in Delay Tolerant Mobile Network", Accepted as an Extended paper to *IEEE International Symposium on a World of Wireless, Mobile and Multimedia Networks (WoWMoM)*, Buffalo, NY, June 2006.

- Khaled A. Harras and Kevin C. Almeroth, "Transport Layer Issues in Delay Tolerant Mobile Network", *International Federation for Information Processing (IFIP) Networking*, Coimbra, Portugal, May 2006.

- Khaled A. Harras, Kevin C. Almeroth and Elizabeth M. Belding, "Delay Tolerant Mobile Networks (DTMNs): Controlled Flooding Schemes in Sparse Mobile Networks", *International Federation for Information Processing (IFIP) Networking*, Waterloo, Canada, May 2005.

Abstract

Protocol and Architectural Challenges in Delay and Disruption Tolerant Networks

Khaled A. Harras

The evolution of wireless devices, such as laptops, personal digital assistants (PDAs), and cell phones, has changed the way networking and communication are perceived. The growing dependence on these devices, along with the high mobility of users, has increased the need to be connected in all places at all times. This increase in user demand, along with the evolution of new applications such as satellite networks, sparse mobile networks, and remote disconnected communities, has ultimately led to the rise of the delay and disruption tolerant networks (DTNs) research area.

The future of computer networks includes an Internet that encompasses numerous heterogeneous networks, reaches out to the most remote areas, and provides communication in the most extreme and unstable conditions. With such a vision, new problems with large complexities arise. Communication devices in the future will be required to remain connected despite the rise of new challenges such as network partitioning, intermittent connectivity, large delays, the high cost of infrastructure deployment, and the absence of an end-to-end path.

This dissertation addresses some of the major problems that arise, as a result of these challenges, in delay and disruption tolerant networks (DTNs). We study several protocol and architectural challenges in DTNs. We specifically introduce and study a new class of DTNs, known as Delay Tolerant Mobile Networks (DTMNs). We propose various controlled flooding schemes in DTMNs, and investigate the performance of these schemes over different DTMN architectures. Additionally, we study the dedicated messenger ownership schemes and scheduling strategies in clustered DTMNs. Furthermore, we address transport layer issues, particularly reliability, in DTN environments. We present a solution for mobile devices experiencing opportunistic intermittent connectivity. Finally, we propose a novel architecture, ParaNets, that encompasses our vision of how future DTN solutions will be built.

Contents

List of Figures

List of Tables

Chapter 1

Introduction

1.1 Motivation

Today's Internet, as well as various other networks, operate on some unstated assumptions, such as small end-to-end Round Trip Time (RTT), the existence of some path between endpoints, having end-to-end reliability with Automatic Repeat reQuest (ARQ), and the use of packet switching as the right abstraction for end-to-end communication. These assumptions have certainly been cornerstones in the design of basic Internet protocols [23], on which the fundamental Internet philosophy [29] and architecture [30] rely and operate.

The evolution of wireless devices such as laptops, personal digital assistants (PDAs), and cell phones, however, has spurred many novel applications with new challenges that oppose such assumptions. The growing dependence on these devices, along with the high mobility of users, has increased the need to be connected

1

in all places at all times. This increase in user demand has spurred the development of numerous new applications. Examples of these applications include: satellite networks, planetary and interplanetary communications, military/tactical networks, disaster response, and other forms of large-scale mobile networks.

These new applications have created a new set of challenges and assumptions for network designers to solve. These new challenges include, but are not limited to, network partitioning, intermittent connectivity, large delays, the high cost of infrastructure deployment, and the absence of end-to-end routes. Network environments characterized by one or more of these challenges are dubbed Delay, or Disruption, Tolerant Networks (DTNs) [34]. Example applications of DTNs include battlefields [61], disaster relief efforts and field hospitals [1], and remote disconnected villages [2], [3], [4].

Some of these new challenges had already spurred much research in mobile environments. Most of the research in this area focuses on that of Mobile Ad hoc NETworks (MANETs), which are multi-hop networks in which nodes cooperate to maintain end-to-end connectivity and perform routing functions within the network. Most of the work in MANETs, therefore, is targeted at solving the routing problem, which has led to the development of various routing protocols [53], [81], [82], [31], [59], [40], [42], [84], [83]. MANET solutions, however, fail to address all of the emerging challenges listed earlier since they focus on scenarios where an end-to-end path *must exist* from a source to a destination. A few attempts consequently emerged to look closer at some of these challenges [97] [68], but these solutions were constrained to very specific scenarios and did not fully address long term common challenges presented by the types of challenged networks with which we are concerned.

The DTN challenges were not fully addressed as a coherent set until the rise of the Delay Tolerant Networking Research Group (DTNRG) [5] which evolved from the Inter-Planetary Networking Research Group (IPNRG) [6]. The DTN community has initially proposed a general architecture [34], [35] that addresses routing issues [52] for networks in extreme environments. Subsequently, many solutions evolved to target DTN-related problems in more specific scenarios and environments. Examples include, but are not limited to, message ferrying [108], [109], [110], sparse sensor networks [55], [88], [14], [89], [99], [102], and different forms of intermittent connectivity [7], [78], [77], [75], [76], [71], [72], [26]. The general DTN architecture works well by trying to address common problems among different networks in challenged environments. Conversely, other specialized solutions are more tailored to the specific problem they try to solve, and therefore, are more customized to their designated environments. Many other problems and issues that were previously overlooked, however, require further research.

These overlooked problems, dissemination of mobile devices, the increased need for challenged network applications, and the alternatives available for wireless connectivity, are all factors that help us build our vision of the networking future. As illustrated in Figure 1.1, we envision an expanding Internet reaching out to remote communities and environments with numerous connectivity challenges, therefore, spurring further research in delay and disruption tolerant networks. As a result of this vision, we set a long term objective to develop and study protocols and architectures that provide us with alternatives to ultimately achieve reliable and robust communication in extreme environments.

Even though the long term objective may seem ambitious and hard to reach, we break it down into a set of manageable challenges that can be addressed by the research community in the short-term. However, with respect to the current state of knowledge and progress in the field, we choose to focus on what we believe to be a crucial subset of these challenges. This subset is shown in Figure 1.1. The first challenge is to design different architectures over which novel solutions can be built and tested in order to determine the best architectures suitable for any given challenged network. The second major challenge has to do with routing alternatives and data delivery techniques in the challenged environments where DTNs are assumed to exist. The last important research challenge concerns the value added services of which the transport services, such as reliability and congestion

4

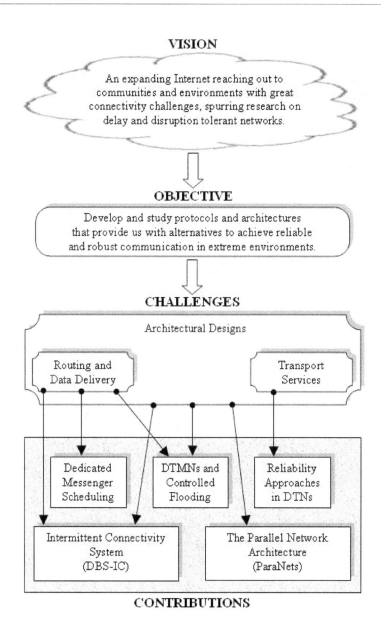

Figure 1.1: Dissertation vision, objective, challenges, and contributions.

control, are most critical. Since architectural design considerations are taken into account when solutions are developed for routing and transport challenges, the architectural design challenge then encompasses the other two challenges as shown in Figure 1.1.

This subset of challenges serves as the motivator for the contributions we present in this dissertation. As the thesis of this dissertation, we posit:

The major challenges in delay and disruption tolerant networks, namely, data delivery, transport services, and architectural designs, are crucial problems to which we propose and study alternative solutions to ultimately achieve reliable and robust communication in these networks.

A summary of our solutions and work presented in this dissertation is shown in the contributions box in Figure 1.1. The following sections in this chapter give a brief overview of these contributions, define their scope, and show the impact of our work. In the end, we discuss the work that remains to complete this dissertation.

1.2   Dissertation Overview and Scope

As depicted in Figure 1.1, our contributions address the set of challenges we believe are the most crucial for delay and disruption tolerant networks. There is another set of challenges that exist in DTNs which we do not address in this dissertation. Examples include multicast, security, and quality of service issues. The work we conduct in this dissertation can be summarized as follows. We introduce and study a new class of DTNs, known as Delay Tolerant Mobile Networks (DTMNs). We study various controlled flooding schemes in DTMNs, investigate messenger scheduling strategies in clustered DTMNs, and examine some transport layer issues in DTN environments. We also present a solution for mobile devices experiencing opportunistic intermittent connectivity. Finally, we propose a vision of future DTN architectures, and how protocols will be built for such networks. We briefly discuss the scope of each of those contributions in this section.

1.2.1 DTMNs and Controlled Flooding

After the general DTN architecture [34] set broad guidelines for thinking about DTN environments, we realized the need for a more specific architecture that targets a subset of DTN applications. As a result, we introduce the Delay Tolerant Mobile Network (DTMN) architecture for sparse delay tolerant mobile networks [45]. Indeed, as DTN architectures evolve and new applications emerge, message flooding remains one of the fundamental communication techniques.

In Chapter 3, we present DTMNs and demonstrate how a variety of controlled message flooding schemes over sparse mobile networks affect message delay and network resource consumption. We examine the use of a probabilistic function for message forwarding. We then add a time-to-live (TTL) or kill time value on top of this probabilistic function and evaluate their performance. Finally, we introduce the idea of a Passive Cure on top of the other schemes and see what effect it has on the network. The Passive Cure is used to "heal" the "infected" nodes, i.e., those that carry a copy of the message. In our work, we study real-life DTMNs in which our work is applicable. We make the assumption that the nodes are totally blind, such that every node knows only information about itself and the messages that it carries. Another important assumption we make is that there is no form of control over any nodes in the network. In other words, each node is completely autonomous and makes its own decisions. These assumptions fit well in several applications where sparse mobile networks exists, such as in disconnected vehicular networks and sensor networks attached to animals for wild life study [55]. We finally study a modified architecture where we assume the existence of high-end stationary nodes to satisfy scenarios where access points or mesh routers might exist in such environments.

1.2.2 Messenger Scheduling

To complement the work presented in Chapter 3, we address message delivery problems that exist in other scenarios such as disaster relief efforts and field hospitals [1], battlefields [61], and remote disconnected villages [2], [4], in Chapter 4.

In these scenarios, we envision a new network environment comprised of *regions* and *messengers*. A region is defined as a cluster of nodes having an end-to-end path between any two nodes in the cluster. Since regions are assumed to be disconnected from each other, we propose using a dedicated set of messengers that relay messages between regions. Each region generates large amounts of data that can be grouped together in *bundles* [34] which are then relayed to other destination regions. The regions could either be mobile, as in search-and-rescue groups or military battalions, or stationary, as in field hospitals or remote disconnected villages. In these environments, we shift the focus from routing [52] and path discovery [110], to *messenger scheduling*. We study how messenger scheduling can be used to improve network performance and connectedness. Furthermore, there are a number of scheduling algorithms (e.g., batching) that can be drawn from marginally related areas and used to provide insight into how best to schedule messengers [43].

In Chapter 4, we develop and study several classes of scheduling algorithms that can be used to achieve inter-regional communication in the environments described above. We introduce the idea of using dedicated messengers under different scheduling schemes rather than discovering and maintaining routing paths in these kinds of networks. We then propose a two-dimensional set of solutions representing different scheduling algorithms for messengers. We describe these algorithms in detail, analyze their performance, and derive their bounds and restrictions. We also examine *adaptive* scheduling strategies to handle sudden behavioral shifts in the system. Our main goals are to gain a better understanding of the challenges involved in the complex network environments we introduce, and to identify which scheduling algorithms are most suitable for different clustered DTNs.

1.2.3 Transport Layer Issues

With the work in DTNs mainly focused on routing, we shift our focus towards studying some transport layer issues in Chapter 5. We particularly study reliability in DTNs, since we believe it to be the most pressing service required for DTNs, and introduce four different end-to-end reliability approaches [44]. First, *hop-by-hop* reliability depends only on sending acknowledgments along every hop in the path. Second, *active receipt* achieves reliability by delivering an *active* end-to-end acknowledgment over the DTMN. Third, *passive receipt* reliability implicitly sends an end-to-end acknowledgment through the network. Fourth, *network-bridged receipt* sends an acknowledgment over another network that exists in parallel to the DTMN. With the multiple devices people currently carry, we can use other parallel networks, such as cell networks, as network bridges to transmit acknowledgements or other control-related information. We evaluate these reliability approaches in DTMNs under various network conditions via simulations. Our goals in Chapter

5 are to examine the impact of these reliability approaches, understand the trade-offs between them, and open the way for further work in transport layer issues in delay tolerant networks.

1.2.4 Intermittent Connectivity

After addressing DTNs where end-to-end connectivity almost never exists, we turn our attention to other delay and disruption tolerant networks that experience a high rate of intermittent connectivity. In such networks, opportunistic connections are often available to nodes to the Internet, rather than to other disconnected nodes. With recent work showing that mobile devices can move at speeds of 75 mph and still experience periods of connectivity with high throughput and low loss [37], we decided to build a system that takes advantage of this fact to better serve mobile nodes experience network disruptions.

In Chapter 6, we present and develop DBS-IC, a Data Bundling System for Intermittent Connections, which takes advantage of short connection periods to enhance the experience of mobile users [48]. In this system, a Stationary Agent (SA), located on a stationary device with a stable connection, collects data the user has specified will be needed in the future. This data can be heterogeneous: data from web servers, email servers, and other file servers. The SA then groups this data together into a single package, or *bundle*. Afterwards, the SA opportunistically sends this bundle to a Mobile Agent (MA), residing on a mobile device, whenever a connection is present. Once the bundle is successfully transferred to the MA, the user can view the data at any time, including times of disconnection. In this way, our system hides the underlying instability of the connection. We fully implement DBS-IC and evaluate its performance in different intermittent connectivity scenarios, and compare the results to existing data retrieval methods. Results of live tests show that DBS-IC efficiently utilizes bandwidth to opportunistically deliver data to the user before disconnections occur. Further details of the system are presented in Chapter 6.

1.2.5 ParaNets

Despite the fact that the challenges posed by delay and disruption tolerant networks have been addressed in different ways for various applications, we observe a general trend in the solutions and architectures presented so far. This trend lies in the fact that all of these solutions are based on the idea of operating over a single homogenous network, or sequential heterogeneous overlay networks.

In parallel to this trend, we have begun to witness the inevitable convergence of different networking technologies. This convergence occurs by providing communication alternatives to users through carrying multiple devices, or a single device, with access to multiple networks [25], [10]. However, network protocols that operate on these devices are primarily designed to operate over a single network at a given time, or multiple channels within the same network. Most, if not all, current approaches for providing inter-operability between heterogeneous networks rely on a high level overlay protocol that performs protocol translation. These overlays, however, are usually at network gateways rather than endpoints. We believe we can exploit the current and expected future convergence of networking technologies to better serve challenged networks.

Based on our vision, as well as the current trends in challenged networks research and networking technologies, we propose the *Parallel Networks (ParaNets)* architecture in Chapter 7. The idea behind parallel networks is to provide an architecture over which network protocols, developed for challenged networks, can seamlessly utilize multiple heterogeneous networks in parallel [46]. Each network can then be used as a channel for the protocol being used. Message types that are best suited for a given network are seamlessly sent using the appropriate channel. We discuss the short-term research challenges and long-term implications of ParaNets in Chapter 7 as well. The work on ParaNets is part of the future work that stems from this dissertation. We ultimately believe that ParaNets will open the door for future protocols capable of providing more robust, timely, and intelligent decisions for challenged networks.

1.3 Contributions

The major contributions of this dissertation can be summarized in the following points:

1. We identify the most challenging obstacles towards realizing a future vision of an Internet reaches remote communities and environments.

2. We introduce a new class of DTNs, known as delay tolerant mobile networks (DTMNs), which specifically focuses on sparse mobile DTNs.

3. We study different flooding schemes over DTMNs showing the pros and cons of each, and examine DTMN architectures where other stationary high-end nodes may coexist.

4. We propose a new means of establishing communication in sparse clustered mobile networks. We use dedicated messengers for such systems and study various messenger ownership schemes and scheduling algorithms.

5. We study and compare different reliability approaches for delay and disruption tolerant networks. We show the tradeoffs between different reliability approaches and costs.

6. We introduce a new system for handling intermittent connectivity that helps mobile nodes take advantage of opportunistic connections that may be repeatedly available.

7. We provide a general architecture for challenged networks, ParaNets, that takes advantage of current networking trends, and offers a solid ground upon which future DTN solutions and protocols can be built.

8. We show the potential impact ParaNets can have on challenged networks by evaluating its performance in a DTMN setting.

Collectively, these contributions present steps towards achieving fully reliable and robust communication in delay and disruption tolerant networks. These contributions also help realize the strengths and limitations of alternative approaches and solutions taken towards achieving our goal. We hope that other researchers in the future will add to and build on the work presented in this dissertation, ultimately bringing us closer to a future where communication can be established regardless of the presented challenges.

1.4 Dissertation Organization

After presenting the motivation, scope and overview, and contributions of this dissertation, we now briefly describe how the remainder of this dissertation is organized. Chapter 2 contains a literature review that sheds some light on the areas related to delay and disruption tolerant networks. Each of the following five chapters correspond to each of our major contributions that we presented in Figure 1.1. We introduce our Delay Tolerant Mobile Network (DTMN) architecture along with the controlled flooding schemes we propose in Chapter 3. We then present the messenger ownership and scheduling schemes, and compare their performance over clustered mobile networks in Chapter 4. Transport layer issues, particularly reliability, is discussed in Chapter 5. Afterwards, we focus on intermittent connectivity and present our Data Bundling System for Intermittent Connectivity (DBS-IC) in Chapter 6. The vision for a futuristic Parallel Networks (ParaNets) architecture over which we believe DTN solutions will be built, is shown in Chapter 7. Finally, we summarize our contributions, highlight the impact of these contributions, and discuss our short term and long term goals and vision, in our conclusions in Chapter 8.

Chapter 2

Literature Review

This section presents a brief overview of the literature and background related to the area of delay and disruption tolerant networks. We first talk about Mobile Ad Hoc Networks (MANETs), describing various routing protocols that were developed to accommodate mobility. We discuss proactive, reactive, and other hybrid approaches to MANET routing protocols. We then shift our focus more on challenged networks, specifically disconnected mobile networks, sparse sensor networks, intermittent connectivity, and other forms of delay and disruption tolerant networks. In the challenged networks section, we shed some light on some of the most prominent contributions that have been made to this area.

2.1 Mobile Ad Hoc Networks

An ad hoc network is a wireless network formed on the fly where no fixed infrastructure exists. Nodes in such networks communicate with each other via other nodes in the network. In other words, each node in the network can act both as an end-host and as a router. Since almost all ad hoc networks are wireless, this implicitly adds the possibility of node mobility. This new challenge contributes to a new set of characteristics specific to mobile ad hoc networks (MANETs) that do not exist in wired networks. Examples include, high error rates, medium contention, and dynamic changes in topologies.

There have been numerous research thrusts in the area of mobile ad hoc networks. A major area that has received a lot of attention in recent years has been that of ad hoc routing, where a variety of routing algorithms and protocols has been presented and developed [12] [50], [31], [80], [53], [81], [82], [40], [41] [83], [42], [12], [59], [60]. These ad hoc routing protocols can be generally classified into two categories, proactive, and reactive. We briefly discuss each of those in the following sections.

2.1.1 Proactive Routing Protocols

Proactive routing, which is similar to routing protocols used in wired networks, generally relies on classic routing tables that are periodically updated. These tables contain entries for all possible destinations in the network. Examples of proactive routing include the Destination Sequence Distance Vector (DSDV) [83], Optimized Link State Routing (OLSR) [50], [31], and the Topology Broadcast based on Reverse-Path Forwarding (TBRPF) [12]. In general, proactive routing works well in networks characterized by their low mobility. Also it allows data packets being generated by the application to be immediately sent, since the route to the destination is already known. These protocols, however, start to break down in networks with high mobility. The reason for this breakdown is that topology changes in such scenarios can be frequent, which will then causes a large number of updates to be generated in order to maintain the routing tables for all the nodes in the network.

2.1.2 Reactive Routing Protocols

Contrary to proactive routing protocols, reactive protocols determine the routes to designated destinations only on-demand. In other words, nodes only search for a route to a destination when there is a need by some application that wants to send data to that destination. Examples of reactive routing include Ad hoc On-demand Distance Vector (AODV) [81], [82] and Dynamic Source Routing (DSR) [53]. Since reactive routing only creates routes on-demand, the amount of control traffic is limited to how active the network is. Stale routes, especially in highly mobile networks, do not exist since routes are created only when needed. The drawback, however, of reactive routing is the initial delay incurred since routes need to be discovered first, before data packets can be sent in the network.

2.1.3 Other Routing Protocols

A mixture of both reactive and proactive approaches is also possible and can be combined with other aspects that may affect the choice of a route, such as geographical information or power awareness. These types of protocols are usually known as hybrid protocols. Examples include Zone Routing Protocol (ZRP) [42] and Location Aided Routing (LAR) [59], [60]. In general, these protocols maintain proactive routes to particular regions, and use reactive routing to other further regions.

While most of the mobile ad hoc routing protocols pick the most appropriate route based on the shortest path between a source and a destination, some protocols use other metrics in order to pick a route. Examples of alternative metrics include link stability [96], load balancing [47], [64], and power consumption awareness [33], [38]. The reason for relying on metrics other than shortest path is to focus more on link stability and reliability. Also, this way more fairness can exist in the network, which ultimately leads to maximizing their lifetimes.

We note in the end of this section, that while most of the work conducted in MANETs address various mobility concerns, they still assume the existence of an end-to-end path from the source node to the destination node. Therefore, these protocols are unlikely to find routes in sparsely connected networks. Our work in this dissertation specifically deals with cases in which this assumption is likely not to be true.

2.2 Challenged Networks

With the growing realization that there are many cases where an end-to-end connection cannot be assumed, several solutions have been proposed that take into consideration the possibility of communication among nodes that are partially or intermittently connected. The term *challenged networks* was first introduced by Fall [34], and continued to evolve to include networks characterized by one or more of the challenges such as large delays, no end-to-end route and high error rates. The notion of challenged networks has become a large umbrella under which a multitude of terminologies have been developed to indicate networks with very similar characteristics. Some of these challenged networking terminologies include eventual connectivity, delay and disruption tolerant, disconnected mobile, intermittently connected, transient connections, partially connected, sparse mobile, opportunistic, and, extreme networks.

In general, all of the solutions for challenged networks rely on some form of store-and-forward approach. The data unit in these networks are defined as messages, large packets, or bundles (several messages grouped together). The evolution of these solutions has been dependent on the spectrum of challenges these solutions attempt to address, as well as the applications they try to serve. We briefly discuss in this section, the different subsets of solutions that have been presented in the area of challenged networks. We note that these subsets are not clear cut subareas within challenged networks, and may easily overlap with each other. We present these subsets as a way to identify different work attempts that have been taken at certain period of time.

2.2.1 Disconnected Mobile Networks

There were early attempts within the mobile ad hoc network community to address networks where disconnections occur. Disconnected mobile ad hoc networks are multi-hop networks where an end-to-end path cannot be assumed to exist between any two end nodes. This fact consequently leads to the failure of all routing protocols that base their algorithms on the assumption of end-to-end connectivity.

An example representing such attempts is that by Vahdat and Becker [97]. They introduce Epidemic Routing, which is a flooding-based routing protocol for partially-connected ad hoc networks. The protocol relies on random pair-wise exchanges of *"summary"* vectors, which is an index of the messages that the host needs to forward and has in its buffer. After exchanging summary vectors, a node sends a message request asking for the messages that it does not have. This is followed by the transmission of these messages. The message eventually propagates through the network and reaches the intended destination. This approach is simple and robust to network partitioning and mobility.

However, epidemic routing produces a large number of redundant messages that can rapidly consume resources (e.g. buffer space and power). This leads to low scalability and high cost. Davis et al. [32] try to minimize the cost incurred in epidemic routing by using node mobility statistics. This way, when the buffer of a given node is full, messages can be dropped in a selective manner based on the probability of meeting other nodes in the future. Similarly, Zhang et. al. present models that reflect the performance of epidemic routing [106].

On the other hand, Li and Rus [68] propose a scheme where nodes actively change or modify their trajectory to help create a path that can more rapidly deliver messages to the destination. This ability is useful in situations where all of the nodes can be controlled and belong to one entity. However, it is difficult to extend this work to support multiple messages.

The work presented in this section was largely regarded as early attempts before the notion of challenged or delay/disruption tolerant networks had been introduced. Further work within this area started adopting the notion of DTNs, and therefore, we discuss in the DTN section later on.

2.2.2 Sparse Sensor Networks

While sparse sensor networks are generally regarded as a form of disconnected mobile networks, we discuss it in this section as a unique application of challenged networks. We focus on attempts that have been made for monitoring or tracking various life forms. The principles behind the challenges faced and solutions spectrums offered, still remain under the umbrella of challenged networks.

Sensor networks is another area where end-to-end connectivity has a high probability of being lost due to sensor mobility. Juang et. al. [55] discuss this problem in the ZebraNet project. In this project, sensors are attached to zebras to monitor their various life patterns. These sensors collect data and forward it when they come into range of a base station. Base stations are mounted on a car or plane that passes within transmission range of the sensors to collect the data.

The sparseness of the deployed sensors also causes the loss of end-to-end paths between nodes. To address this problem, Shah et al. [88] introduce DataMULEs, where low powered static sensors are sparsely deployed to gather various forms of data, and then a mobile entity, a "mule", travels among these sensors to collect the data they gathered. Similarly, there are many other examples that involve comparable scenarios where sensors need to be attached to seals [14] or whales [89] in the ocean in order to collect various oceanographic data.

2.2.3   Intermittent Connectivity

Another area that falls under challenged networks, has to do with intermittent connectivity. In these networks, nodes experience a high rate of short disconnections, as opposed to opportunistic short connections with large disconnection times. Two main methods exist to counteract the detrimental effects of these short disconnections. The first approach involves maintaining session-level connections through disconnections [7], [8]. Ott and Kutscher first examine the feasibility of mobile network traffic for in-motion users [75]. They introduce their Drive-thru Internet Architecture and examine a Connectivity-Loss Resilient Connection (CLRC) between a mobile client and a fixed proxy that maintains information regarding multiple TCP streams [76]. By maintaining the CLRC and splitting the connection at a proxy, transport-level connections remain open through disconnections. Mao et al. present a similar approach, maintaining session-level connections through disconnections [71]. The goal here is to allow the user to seamlessly resume applications upon return of connectivity. Comparably, Kulkarni et al. discuss methods to keep an unreliably connected mobile client synchronized with rapidly changing web page content [62]. Their solution uses a proxy that sits between the client and server, and caches requests during times of disconnection.

The second approach to intermittent connectivity does not try to maintain high-level connections, but simply delays delivery of data while the mobile device is disconnected. This approach includes solutions such as middleware to 'store-and-forward' client requests during times of disconnection or weak signal [57] and to synchronize data once connectivity returns [72]. Similarly, Chang et al. present an ARTour Express program which stores requests internally so the user can seek multiple pages without waiting for each to completely load [26].

2.2.4 Delay and Disruption Tolerant Networks (DTNs)

Within the Internet Research Task Force (IRTF), the Delay Tolerant Networking Research Group (DTNRG)[5] evolved from the Inter-Planetary Networking Research Group (IPNRG)[6] that initially developed an architecture for inter-planetary networks [49], [24], [18]. Fall provides a generalized Delay Tolerant Network overlay architecture as an attempt to achieve inter-operability between various heterogeneous networks deployed in extreme environments [34]. Such networks often lack continuous connectivity and suffer from potentially long delays. A *bundle layer protocol* is introduced to handle many of the challenges previously discussed using a store-and-forward approach, or through custody transfer, where a *custodian* assumes the responsibility of reliably delivering a bundle to the next custodian on the path to the destination [35]. Jain el al. expand on the DTN work by studying routing issues in such extreme environments [52]. Further work was also conducted by the same authors to cope with failures in DTNs [51]. After these initial attempts to explore and study problems in DTNs, research in the area drastically grew with dedicated workshops and conferences emerging to reflect this growth. Researchers that were using different names prior to the introduction of DTNs gradually migrated towards adopting the DTN notion, as well as that of challenged networks for the work [9].

While the work conducted in the area of DTNs has currently grown to include many different thrusts, most of the attention has been given to routing [107]. While there are a few deterministic approaches to DTN routing based on different knowledge oracles [52], most approaches rely on some form of randomization or stochastic approaches. Examples include epidemic or random approaches [97], [39], [45], [58] [89], [74], [91], [94], history or prediction-based approaches [32], [70], [73], [95], [54], controlled mobility [68], [109], [110], [43], and coding-based approaches [51], [100], [103], [69]. Numerous other routing approaches have also been proposed for DTNs [58], [17], [67], [27], [79], [105], [11], [65], [66], [20], [21], [22]. We have already discussed some of these approaches in the previous sections, and now take a brief look at several other routing approaches.

Considering random schemes for DTN routing, an extreme approach to epidemic routing [97] is for a node to hold the message until it delivers it directly to the destination when it comes within communication range [39]. Furthermore, the Mobile Relay Protocol (MRP) integrates message routing and storage in the network [74]. When using MRP, the source node broadcasts the message to its immediate neighbors, which then enter into a relay mode trying to forward packets through a route that may exist to the destination. If no route exists, the neighbors enter a storage phase until the destination comes within range.

Aside from the epidemic approaches, solutions based on the likelihood of the delivery of a neighbor have been proposed. Davis et. al. extend the work in epidemic routing [97] to scenarios where resources are limited, with a major focus on message dropping strategies when buffers are limited [32]. PROPHET (Probabilistic Routing Protocol using History of Encounters and Transitivity), on the other hand, estimates a probabilistic metric known as the delivery predictability to decide whether a message should be forwarded it or not [70]. Similar to PROPHET, Context Aware Routing (CAR) adapts its routing strategy between synchronous and asynchronous mechanisms for message delivery to handle intermittent connections in mobile networks [73].

Other DTN routing protocols base the forwarding decision on end-to-end performances. The meets and visits (MV) protocol, for example, learns the frequency of node encounters and their visits to certain regions [20]. The shortest expected path routing (SEPR) protocol, on the other hand, estimates the link forwarding probability based on previous data stored in each node [95]. Similar to SEPR, the minimal estimated expected delay (MEED) protocol focuses on the contact summary oracle discussed in Jain et. al. [52]. Overall, MEED computes the expected delay to deliver a certain message, based on history records of connection and disconnection times between contacts in the network [54]. All these approaches, in general, perform better than epidemic routing, with an increased cost in complexity.

Node movement control, is another approach towards solving the routing problem in DTNs. A similar example to the work we proposed by Li and Rus [68] in the previous section is that of the Message Ferrying (MF) [108]. Authors of MF basically propose controlling the movement of ferries to gather data from stationary nodes in a sparse network [108]. They expanded their work with Zegura to consider networks with mobile nodes [109]. They also built on this work to explore controlling the mobility of multiple ferries by studying different algorithms that compute different routes for the ferries [110]. Another approach to node movement control approaches is that proposed by Burns et. al. where autonomous mobile agents are added to a network in order to increase the overall performance [20], [21], [22]. The authors study different agent control algorithms to optimize the network performance based on a quality metric chosen by the network administrator.

Other research thrusts in DTNs have recently shifted focus from the routing issue to include several other areas. Transport layer services such as congestion control [86], [19], and reliability approaches [44], and addressing breaks in TCP [36], have been visited. Security issues [87], [16], multicast [111], and various tradeoffs [90], [56], [101], [103] have also been given some attention. Developing further applications for developing regions has been the focus of several research groups as well [2], [15], [3]. Finally, novel architectures like Haggle [85], ParaNets [46], EDIFY [28], and the Internet Indirection Infrastructure [93], [112], [63], have also been studied and proposed.

We believe that all these new thrusts in DTNs open the way for future research opportunities and challenges. We point out that this section in the dissertation will grow to be a contribution in itself. We plan to carry a full survey study where we can classify this work in a way that would help future researchers indicate commonalities within the different sub areas in DTNs, as well as identify possible future research routes that may be adopted.

Chapter 3

Controlled Flooding in Sparse Mobile Networks

3.1 Introduction and Motivation

As challenged network architectures evolve and as new applications emerge, message flooding remains one of the core communication techniques. In this chapter, we demonstrate how a variety of controlled message flooding schemes over sparse mobile networks affect message delay and network resource consumption.

We introduce a general architecture for sparse mobile networks known as Delay Tolerant Mobile Networks (DTMNs). We propose and study several flooding techniques over DTMNs, which include the use of a probabilistic function for message forwarding. We then add a time-to-live (TTL) or kill time value on top of the probabilistic function and examine the reduction in cost. Finally, we add our novel idea of a Passive Cure on top of the other schemes and see what effect it has on the network. The Passive Cure is used to "heal" the "infected" nodes, i.e., those that carry a copy of the message.

We study real-life sparse mobile networks in which our work is applicable. In such cases, we assume that the nodes are totally blind, such that every node knows only information about itself and the messages that it carries. Another important assumption we make is that there is no form of control over any nodes in the network. In other words, each node is completely autonomous and makes its own decisions. To get closer to real networks, we present an alternative architecture that includes high-end stationary nodes which represent the expected widespread deployment of access points or mesh routers. In the end, we extensively evaluate our system along with the proposed controlled flooding solutions via simulations.

3.2 Delay Tolerant Mobile Networks (DTMNs)

This section describes our proposed architecture. We first describe the components of our architecture, along with the functionality and behavior of each component. We then introduce the notion of *"willingness"*, which is a reflection of the degree that nodes are willing to participate in relaying messages. Following that description, we discuss general strategies for controlling message flooding over sparse mobile networks. We then define the specific control schemes that we study in this chapter. This description is followed by algorithms depicting the system operation. Finally, we present our extended architecture, an extension that incorporates high-end nodes.

3.2.1 Architectural Components and Assumptions

To satisfy the problems and challenges posed by the scenario offered in Section I, we propose a transport layer overlay architecture for sparse mobile networks. Message forwarding and handling is done by this overlay layer and handles the heterogeneous protocols and node characteristics. This approach also complies with the basic DTN architecture proposed by Fall [34].

There are two important assumptions we make in our system. The first is *node blindness*. Nodes in the network do not know any information regarding the state, location or mobility patterns of other nodes. The second is *node autonomy*. Each node has independent control over itself and its movement. The reason behind these assumptions is to closely model the real world scenarios described earlier. When driving through a sparse environment, any given node has no knowledge of other nodes that come within its range (Blindness), and it is autonomous in its movement (Autonomy).

In Figure 3.1, we show the nodes in the network divided into three types. A *sender node, "S"*, is the node that initiates the transmission of a message to a destination in the network. A *forwarder node*, *"F"*, is any node that carries the message from the sender, or another forwarder, with the aim of relaying it to the ultimate node. Finally, the *ultimate node*, *"U"*, is the final destination.

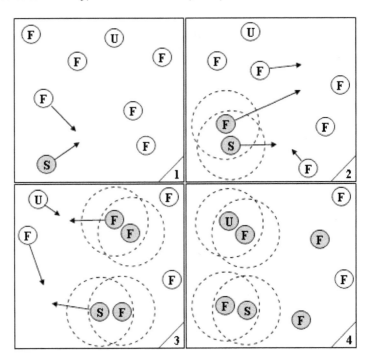

Figure 3.1: The DTMN architecture.

The basic mechanism of node interaction is shown in Figure 3.1. The interaction of nodes is similar to that in Epidemic Routing [97], where each node continuously tries to relay a message to other nodes, within range, that do not already have the message. We look at an example where a sender node, S, needs to send a message.

In the beginning, S initiates a periodic beacon for neighbor discovery purposes. It announces that it has a message that needs to be forwarded to a specified destination, U. When S comes within range of one or more forwarder nodes, F, or even the ultimate node, U, the beacon is received and an ack is sent to S from each node that received the beacon and does not have a copy of the message. When S receives an ack for its beacons, it simply broadcasts the message to its neighbors. Once the message is received, the forwarder node starts to act as a sender node. It sends its own beacons, and both nodes travel through the network looking for either another forwarder to pass the message on, or for the ultimate node. The message gradually propagates through the network until it eventually reaches the ultimate node. This process results in the overuse of network resources through continuous and repetitive flooding of messages. On the other hand, the advantage of this approach is the high delivery rate and relatively small delay.

3.2.2 Modeling Node Willingness

Generally speaking, the frequency at which a sender or forwarder node tries to forward a message depends on many factors. Some of these factors include node state (power or buffer space, for instance) or message state (size or priority of message). We model this as the *willingness* of a node. The willingness of a node is the degree to which a node actively engages in trying to re-transmit a message. Willingness can be modeled in terms of three variables. First, the *beacon interval* is the amount of time a sender or forwarder node waits before sending a new beacon. Second, the *times-to-send* is the number of times a node successfully forwards a message to other nodes in the network before it chooses to stop forwarding the message. Third, the *retransmission wait time* is the amount of time a node waits without beaconing before it tries to resend the message to other nodes in the network. The source node includes the value of these parameters as part of the message header. This way, the forwarder nodes can set their willingness levels accordingly.

To help clarify how these three variables affect the behavior of a given node in the network, we introduce the following simple example. Let us assume that the beacon interval = 1 sec, times-to-send = 2, and the retransmission wait time = 50 sec. These values indicate that when a sender wants to send a message, it sends a beacon every second to find other nodes that would carry the message. Once the sender node finds a forwarder node, by receiving an ack for its beacon, it transmits the message and decrements the times-to-send by one. The sender then waits for 50 seconds before it resumes sending beacons every second to look for the next node to which it will forward the message. This process repeats until the times-to-send reaches zero. The forwarder nodes that received the message in both cases start acting as the original sender, assuming that all nodes have the same willingness.

3.2.3 General Controlled Flooding Strategies

Despite the varying willingness levels of nodes, there still exists a problem of network resource consumption. In the basic flooding case, there are numerous copies of the same message propagating throughout the network, utilizing resources (e.g., buffer space, power, and bandwidth) in order to deliver a single message. To control this flooding, we identify several "stopping" strategies where message flooding is controlled through the decision of the nodes to stop forwarding the message. These strategies satisfy our assumptions of node blindness and autonomy; if those assumptions are relaxed, other more optimized techniques could then be used. We categorize the various stopping techniques to control message flooding into the following categories:

Node-State: Here, a node decides whether to forward the message based on several conditions related to its own state. One example includes the amount of power available in the node (for instance, a PDA connected to the car battery could forward a message more than a low powered laptop). The decision can also be based on the amount of buffer or storage space the node needs to reserve for the message while it continues forwarding it. A node's mobility can also affect its decision to forward a message. A node that is stationary might not try to frequently send the message, since its probability of meeting other nodes is unlikely to be less than a highly mobile node (unless that node is in a hot spot, for instance).

Message-State: In this case, a node's decision to forward a message is based on attributes of the message itself. One common technique is that of a hop limit, or time to live. Each message has a maximum time during which it can be forwarded, after which, it is dropped. Another technique is to use message priority or size. Messages with higher priority replace those with a lower priority if the buffer is full. Alternatively, smaller messages can be forwarded before larger ones if, for example, a node that cannot carry a large message could carry a smaller one.

Environment-State: The decision here is based on surrounding environmental factors. For instance, a node may decrease the number of times it sends the message if it realizes that it is in a dense environment. This knowledge could be sensed either by the number of acks the node receives for its beacons, or simply by receiving beacons from may other nodes. Another factor may be having too much noise or interference which might make the node try less if the message is not of high priority, for instance.

3.2.4   Specific Controlled Flooding Schemes

We study different controlled flooding schemes. We are careful to base our schemes on simple, non-chatty, and elegant algorithms. We choose this goal, because, in sparse and highly mobile networks, complex or chatty algorithms waste the short time nodes have when they come within range of each other. The schemes we introduce are the following:

1) Basic Probabilistic (BP): When describing the *willingness* of a node in the previous section, we implied that forwarder nodes have the same willingness as the sender node. To more closely emulate reality, however, we choose a uniform probability distribution that determines the willingness of the nodes to transmit a given message. Based on the result of this function, a forwarder may choose not to forward the message at all, forward it at half the willingness of the sender, or forward it at the same level of willingness as the sender.

2) Time-to-Live (TTL): In this scheme, we add a time-to-live value. The TTL here determines how many times the message is forwarded before it is discarded. We add the TTL on top of the BP scheme since the BP scheme is a more realistic representation of how nodes act regarding the choice to forward messages.

3) Kill Time: Here, we add a timestamp to the message on top of the BP scheme. The timestamp is the point in time after which the message should no longer be forwarded. The timestamp is an absolute universal life time for the message. This technique would be appropriate if, for example, the sender node knows how long it will be disconnected. This technique is also a good way to set the maximum time a node should keep a message in its buffer if the times-to-send (TTS) variable of that message does not reach zero.

4) Passive Cure: The final scheme, or optimization, we introduce is a Passive Cure. The idea is that, once the ultimate node receives the message, it generates a Passive Cure to "heal" the nodes in the network after they have been "infected" by the message. The ultimate node "cures" the forwarder that passed the message to it by sending a "cure-ACK" instead of an ordinary ACK that is sent when a beacon is received. Whenever a forwarder or the ultimate node detects any other node sending the same message, it sends a cure-ACK to that node to prevent future retransmissions.

3.2.5 System Operation

In this section, we offer a more detailed description of our system's operation of our system as well as the specific control flooding schemes we are studying. We do so by presenting the algorithms that represent the operation of sending a single message from a source node to the ultimate node in a sparse mobile network. These algorithms focus on the operation of the forwarder nodes, since their operation is the determining factor for how the system operates as a whole.

We assume that the passive cure is applied to the system along with the other stopping techniques. The algorithms are shown in Figures 3.2, 3.3, 3.4, and 3.5.

Figure 3.2 describes the general notation that will be used for all the algorithms in this section (Lines 1-4), as well as the forwarder node setup and initialization (Lines 5-8). All forwarder nodes begin as "uninfected" with respect to the message that needs to be delivered (Line 7). The willingness level of each forwarder is then set according to the probability function we introduced earlier (Line 8). Meanwhile, the source is now trying to send a message to the ultimate destination.

Figure 3.3 depicts the operation of an uninfected forwarder. As long as the node is uninfected, it roams around until it receives a beacon from the source or another infected forwarder (Lines 9-11). If the willingness of this forwarder is not zero, it acknowledges the beacon and receives the message (Lines 11-13). If the TTL value on the received message reaches zero, the message is then dropped

```
1: // Notation
2: S: Sender, U: Destination, F: Forwarder, B: Beacon
3: m: Message, n: Total # of Nodes, W: Willingness
4: →: Sends, ←: Receives, tts: Times-To-Send

5: // Forwarder nodes setup and initialization
6: for every F_i where i goes from 1 to n-2 do
7:     F_i.state_m = Uninfected;
8:     W_{F_i} = f_{probabilistic}(n);
```

Figure 3.2: Notation and system startup.

```
9: while (F_i.state_m == Uninfected) do
10:    F_i moves according to mobility model;
11:    if ((F_i ← B_m from F_j or S) && (W_{F_i} ≠ 0)) then
12:        F_i → ACK_{B_m} to F_j or S;
13:        F_i ← m from F_j or S;
14:        if (− − m_{TTL} == 0) then drop m and continue;
15:        F_i.state_m = Infected;
16:        F_i.tts_m = W_{F_i}(m);
17:        if (Kill_Time == True) then
18:            Update Kill_Time_m accordingly;
```

Figure 3.3: Uninfected forwarder node operation.

and the forwarder continues to move without being infected (Line 14). Assuming the TTL is not zero, the forwarder is then infected with the message, sets the number of times it needs to forward the message (the TTS value) according to its willingness level, and updates the message kill time (Lines 15-18).

Figure 3.4 illustrates the operation of an infected forwarder. In general, the infected forwarder keeps moving and tries to pass the message to other uninfected forwarder nodes, or to the ultimate node as long as the kill time and TTS values for the message do not reach zero (Lines 18-20). Once the infected forwarder receives an acknowledgement for its beacon from an uninfected forwarder, it forwards the message and updates the TTS value (Lines 21-22). If the uninfected node happens to be the ultimate node, the infected node then receives a cure acknowledgement and sets its state to cured (Lines 23-26). Meanwhile, the infected forwarder ignores any beacons it might receive for the message it is currently trying to send, and occasionally updates the message kill time (Line 27-28). Finally, if the while loop ended without the forwarder being cured, the message is dropped and the forwarder returns to the uninfected state.

$$
\begin{aligned}
&18:\ \textbf{while}\ ((F_i.state_m == \text{Infected})\ \&\& \\
&19:\qquad (F_i.tts_m > 0)\ \&\&\ (\text{Kill\_Time}_m > 0))\ \textbf{do} \\
&20:\quad F_i\ \text{moves and broadcasts}\ B_m; \\
&21:\quad \textbf{if}\ (F_i \leftarrow \text{ACK}_{B_m}\ \text{from}\ F_j)\ \textbf{then} \\
&22:\qquad F_i \rightarrow m\ \text{to}\ F_j;\quad F_i.tts_m = F_i.tts_m - 1; \\
&23:\quad \textbf{if}\ (F_i \leftarrow \text{ACK}_{B_m}\ \text{from}\ U)\ \textbf{then} \\
&24:\qquad F_i \rightarrow m\ \text{to}\ U; \\
&25:\qquad F_i \leftarrow \text{ACK}_{Cure}\ \text{from}\ U; \\
&26:\qquad F_i.state_m = \text{Cured};\quad F_i.tts_m = 0; \\
&27:\quad \textbf{if}\ (F_i \leftarrow B_m\ \text{from}\ F_j)\ \textbf{then ignore}; \\
&28:\quad \text{update Kill\_Time}_m; \\
&29:\ //\ \text{We examine why the while loop ended} \\
&30:\ \textbf{if}\ (((F_i.tts_m == 0)\ \|\ (\text{Kill\_Time}_m == 0))\ \&\& \\
&31:\qquad (F_i.state_m \neq \text{Cured}))\ \textbf{then}\ F_i.state_m = \text{Uninfected};
\end{aligned}
$$

Figure 3.4: Infected forwarder node operation.

The last state in which a forwarder node can operate, the cured state, is shown in Figure 3.5. The forwarder in this case basically moves around waiting for other infected nodes to initiate contact with it, so long as the cure time does not expire (Lines 32-33). The cure time is a timer chosen sufficiently large, after which the cured state can be deleted by the node. Once the cured forwarder receives a beacon from an infected forwarder or the source, it sends its cure acknowledgement in order to heal the infected node, and updates its cure time (Lines 34-36). Once the cure time expires, the forwarder returns to its initial uninfected state (Line 37).

32: **while** $((F_i.state_m ==$ Cured$)$ && (Cure_Time $\neq 0))$ **do**
33:     F_i moves according to mobility model;
34:     **if** $(F_i \leftarrow B_m$ from F_j or $S)$ **then**
35:         $F_i \rightarrow$ ACK$_{Cure}$ to F_j or U;
36:         update Cure_Time;
37: **if** (Cure_Time $== 0)$ **then** $F_i.state_m =$ Uninfected;

Figure 3.5: Cured forwarder node operation.

3.2.6 Extended Architecture with H-nodes

The expected deployment of access points and mesh routers, along with a study showing that mobile devices can experience periods of connectivity with high throughput, even at speeds of 75 mph [37], leads us to consider a modified architecture. We extend our basic architecture to include what we call "H-nodes". H-nodes are basically high-end nodes fully connected to each other through wired or wireless backbones. These H-nodes represent deployed access points or mesh routers that may exist in some portions of a sparse mobile network. H-nodes are therefore stationary nodes characterized by having higher transmit power and storage capacities.

After studying various controlled flooding schemes, we choose the most promising combination of schemes that give the best performance based on the results shown in Section IV. We use this scheme in our extended architecture and study the impact of having these high-end nodes in our system. Figure 3.6 demonstrates how the extended architecture operates. The figure is based on using a combination of Passive Cure + TTL + BP to control message flood in the network. This combination is what has proven to work best according to our simulation results.

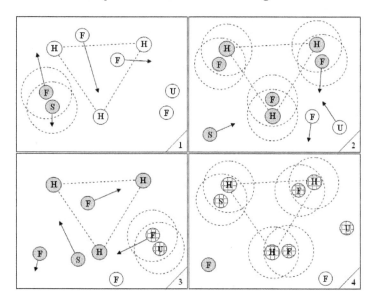

Figure 3.6: Extended architecture with H-nodes.

Figure 3.6 shows three H-nodes deployed and fully connected to each other. Other mobile nodes move in the network. When S has a message to send, it passes it to the first forwarder node that comes within range. When F comes within range of any H-node, all the interconnected H-nodes are infected by the message. These

H-nodes consequently infect other forwarder nodes that are within their range. Once U receives the message, it sends its "cure" to heal the forwarder that gave it the message. The cured forwarder then moves towards one of the H-nodes thereby curing it along with all the other H-nodes and F nodes within their range. In this new architecture, the ultimate destination may be one of the H-nodes. This condition occurs, because, in many cases, the email or transaction that needs to be sent, simply needs an access point or mesh router connected to the Internet. If U is a node in the network, then the H-nodes will help deliver the message faster, as shown in Figure 3.6.

In general, if U is not one of the H-nodes, then the H-nodes could act in several different ways based on their unique characteristics. First, they could actively forward the message onto other forwarders, thus, acting the same way a forwarder node would act, but with full willingness. Second, the H-nodes could initiate the Passive Cure, and be responsible for passing the message only when the U node comes within its range. The advantage of this technique is that fewer nodes will be infected, but the delay might be higher. Finally, they could use high power signals, in either of the previous two modes, to send the message over larger ranges.

The goal of introducing this extended architecture is to demonstrate how a simple disconnected sparse mobile network could evolve to be a more realistic combination of ad hoc and/or mesh network, with potentially numerous Internet gateways. In any of these architectures, our algorithms are designed to be efficient, effective, and elegant.

3.3 Evaluation

In our evaluation, we seek to achieve three main goals. First, we hope to analyze how a controlled flooding scheme behaves in general. Second, we want to measure how the specific flooding schemes affect network efficiency and overall delay. Finally, we examine the impact of the enhanced H-node architecture on our metrics. We now describe our simulation setup and environment and follow it by summarizing the outcomes of an extended set of simulations.

3.3.1 Simulation Environment

We conducted our simulations using the GloMoSim network simulator [104]. We added an overlay layer that handles all the message bundling and relaying, and implements the controlled flooding schemes that we have described. Since we do not have real movement data for our target scenarios, we use a *modified* random way-point mobility model that avoids the major problem of node slow down in the conventional random way-point model. Even though there are other more realistic mobility models that have been developed, we believe that they do not accurately model the scenarios we are concerned with. Therefore, we adopt the simple random way-point model since we believe as well that the mobility model will have little impact on the "relative" performance of our schemes. The node speed ranges between 20 to 35 meters per second; and the rest period is between 0 and 10 seconds. We note that we studied other slower speeds, but chose to show results with these high speeds to show that our approaches work even at these high-end cases. Finally, every point in our results is taken as an average of ten different seeds.

The major parameters used in our simulations are summarized in Table 3.1. The *terrain* is the area over which the *number of nodes* are scattered. Each node has a *transmission range* of 250m. *Simulated time* represents the amount of time the simulations run. The *beacon interval* is the period after which beacons are sent. A "beacon" is simply a signal emitted by all nodes to search for other nodes in the network as well as to announce its location. The *times-to-send* (TTS) is the number of times a node will successfully forward a message to other nodes in the network. *Retransmission wait time* represents the amount of time a node remains idle after successfully forwarding a message to another node. When the retransmission wait time expires, the node then tries to resend the same message. We mainly use TTS to represent the willingness of the nodes to participate in

message relaying. The *time-to-live* is the number of hops after which a message is
discarded. *Kill time* is the amount of time after which the message is discarded.
Finally, the *number of H-nodes* are the number of high-end nodes scattered in the
system.

Table 3.1: DTMN controlled flooding simulation parameters

Parameter	Value Range	Nominal Value
Number of Nodes	25 to 200	100
Terrain	$10km^2$ to $50km^2$	$10km^2$
Simulated Time	1hour to 24 hours	6 hours
Transmission Range	250m	250m
Beacon Interval	0.5sec to 50sec	1sec
Times-To-Send (TTS)	1 to 50	10
Retransmission Wait Time (RWT)	0sec to 500sec	50sec
Time-To-Live (TTL)	2 to 10	7
Kill Time	1000sec to 21600sec	5000sec
Number of H-Nodes	1 to 10	1

We consider two main metrics in evaluating our controlled flooding schemes.
First, *network efficiency* is represented by the total number of messages sent by
the nodes in the network. Second, *overall delay* is the total time that elapses
from when a node wants to send a message until the ultimate node receives that
message for the first time.

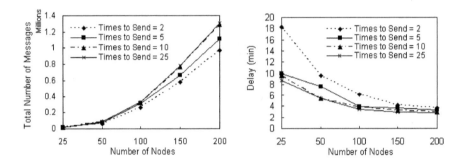

Figure 3.7: The impact of changing the number of nodes and times-to-send (TTS) on the (a) total number of messages, and (b) overall delay in the basic scheme.

3.3.2 Simulation Results

Before applying our probabilistic scheme, we analyze how the network acts assuming full willingness of all the nodes in the network along with enforcing some basic level of message control. This scheme is very similar to the Epidemic Routing scheme, but with some control over message forwarding. We use the times-to-send (TTS) variable to represent node willingness. We investigate how varying the TTS value and the network density affect our metrics. The retransmission wait time (RWT) in these experiments is 1 sec.

Figure 3.7(a) shows the total number of messages sent by all the nodes in the network and Figure 3.7(b) shows the delay. Overall, Figure 3.7 shows that increasing the density of the network results in a large increase in the total number of messages sent. One interesting point is that an increase in the network density results in a significant decrease in the overall delay only up to a certain point (number of nodes = 100), after which, the decrease in delay is overshadowed by the corresponding increase in cost. On the other hand, increasing the willingness beyond a certain point (TTS = 10) does not have a large effect on either metric. The reason for this result is that the nodes in such sparse networks do not encounter each other often enough to consume the large value of the times-to-send (TTS).

Basic Probabilistic Behavior

After applying our basic probabilistic scheme, we examine how the network density and retransmission wait time (RWT) impacts our metrics. We assume that 25% of the nodes in the network have zero willingness, 25% have full willingness, and 50% of the nodes forward the message with only half the willingness (i.e., half the TTS of the source node).

Figure 3.8 shows the result of varying the network density while keeping the TTS set to 10. One interesting observation is the significant drop in terms of the total number of messages when the RWT increases, with only a corresponding small increase in delay (until RWT reaches 100). With a small RWT, the message spreads rapidly through the network, and in a very short time, many forwarders are actively trying to send the message. As the RWT increases, the message does

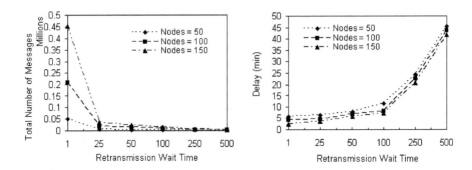

Figure 3.8: The impact of changing the retransmission wait time (RWT) and number of nodes on (a) the total number of messages, and (b) overall delay.

not initially reach as many forwarders. This behavior results in a significant drop in the number of messages. Also, with a large RWT, the TTS is not consumed as quickly, so a forwarder node acts as a forwarder longer, thus rejecting receipt of the same message. With a small RWT, the TTS is quickly consumed. Because the nodes do not keep any state or cache, they are ready to receive the same message and start forwarding it again.

Similar result patterns are generated when keeping the number of nodes constant, but while varying the TTS. We have also seen similar patterns when we used only the basic controlled scheme, but with a much larger scale in terms of the total number of messages. This result shows how a uniform probability based scheme performs much better. The results are not shown due to space limitation.

Time-to-Live and Kill Time

Keeping the basic probabilistic scheme, we build the other schemes on top of it to see their impact on our metrics. Figure 3.9(a) shows the impact of adding TTL only and adding TTL + Passive Cure, to the basic probabilistic scheme. We discuss the Passive Cure results later, and here focus on the TTL.

The first result we observe in Figure 3.9(a) is the large decrease in the total number of messages sent in the network (note the scale is significantly smaller than in Figures 3.7(a) and 3.8(a). This result occurs because the TTL puts a limit on the number of times a message is forwarded. Previously, messages endlessly propagated throughout the network with no limit other than the nodes' willingness.

The impact of TTL on delay is not show here. However, we found that as the TTL increases, the overall delay decreases up to a certain point (at TTL = 7), after which it remains relatively constant at 10 minutes. Even though there is a probability of message loss when using low TTLs, we believe it is a better choice when added to the basic probabilistic scheme. If a large TTL is set, our simulations show that all messages reach the destination, and the cost is much smaller than in the BP scheme. For instance, if we choose a TTL of 9, we incur a cost of 1000 messages instead of 25000 messages, while keeping the overall delay the same for both cases.

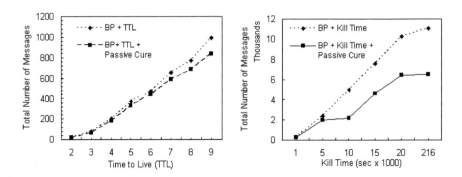

Figure 3.9: The impact of adding (a) TTL and Passive Cure to BP, and (b) Kill Time and Passive Cure to BP.

Figure 3.9(b) shows the impact of adding only the kill time mechanism, and adding kill time + Passive Cure, to the basic probabilistic (BP) scheme. The kill time scheme introduces another improvement over the basic probabilistic scheme alone. The use of a universal time after which a message is discarded certainly stops message transmission and propagation. Figure 3.9(b) shows how the total number of messages increases as the kill time is increased. On the other hand, the overall delay (not shown due to space limitation) remains constant at 10 mins. The only disadvantage of the kill time is if it is set too small so that the message does not make it to the destination.

Passive Cure

In Figures 3.9(a) and 3.9(b), we also demonstrate the effect of adding the Passive Cure optimization to the TTL and kill time, respectively. The Passive Cure does not introduce any improvement in terms of delay because it does not help the message get to the ultimate node any faster. The effect of the cure is evident only in the total number of messages sent in the network. As Figure 3.9 shows, when the Passive Cure is added, there is a drop in the total number of messages. This drop is due to the fact that when the cure starts to operate, the cured nodes stop trying to forward the messages even if the kill time has not been reached or if the TTL has not been consumed.

The Passive Cure optimization has several advantages over the other schemes. First, if the message reaches the ultimate node early, little network flooding may take place since the cure stops the flood early on. Second, the ultimate node receives the message only once. In all the other schemes, however, the ultimate node may receive the same message multiple times. Finally, the Passive Cure may be used as a way to implement end-to-end acknowledgments if it is forced to propagate back to the sender node, since the sender would then know that its message reached the ultimate node.

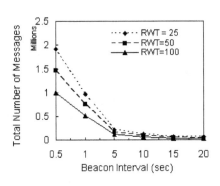 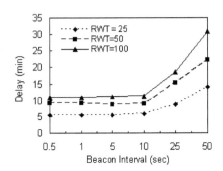

Figure 3.10: The impact of the beacon interval on (a) the total number of beacons, and (b) the overall delay.

Beacon Interval

In order to perform all the message exchanges in a sparse mobile network, neighbor discovery is of great importance. A node occasionally broadcasts beacons in order to announce its location and search for other nodes that may be within its range. The frequency of these beacons has a great impact on the metrics we consider in our system. This impact is illustrated in Figure 10, where we show results for the impact of the beacon interval time on the number of beacons sent and the total delay.

The overall results in Figure 3.10 show, in general, that smaller beacon intervals result in better neighbor discovery which leads to a larger number of messages (not shown) and beacons being sent. This result leads to a faster spread of the message in the network which results in a lower overall delay. The interesting observation, however, is the fact that a small change in the beacon interval can lead to a large decrease in cost with minimal impact on delay. This fact is clearly shown for the beacon interval between 0 and 10 seconds. We can see a large drop in terms of the number of beacons (Figure 3.10(a)), with almost no increase in the average delay (Figure 3.10(b)). The reason for this behavior is that, for any given mobile system, depending on the mobility and density of the network, there will be a threshold in terms of beacon interval time, below which no further gain can be achieved.

Scheme Comparison

In this section, we compare the performance of all the flooding schemes. Overall, Figure 3.11 shows the performance of these schemes relative to each other. When looking at Figures 3.11(a) and 3.11(c), we observe that the basic probabilistic scheme by itself is the most expensive, while the basic probabilistic + TTL + Passive Cure is the least expensive in terms of the total number of messages and the total number of beacons sent. Note that the basic probabilistic scheme already performs better than the basic flooding technique, which is analogous to Epidemic Routing.

Figure 3.11(b) shows that most of the schemes have similar overall delay. The advantage is for the basic probabilistic and basic probabilistic + Passive Cure

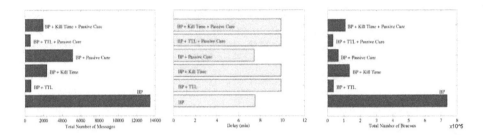

Figure 3.11: Comparing the impact of the controlled flooding schemes on (a) the total number of messages, (b) the overall delay, and (c) the total number of beacons.

schemes. The reason for this result is that the complexity introduced by the TTL or the kill time results in an overall increase in delay. However, a two minute increase, with a corresponding large decrease in number of messages and beacons sent, certainly seems acceptable.

Figure 3.11 summarizes the general tradeoffs between the schemes we introduce. For example, adding the Passive Cure to the basic probabilistic scheme saves about 60% of the total number of messages, with no increase in the overall delay. When adding the TTL to the basic probabilistic scheme, the total number of messages is reduced by more than 90%, while the overall delay increased by only two minutes. This increase in delay does not notably affect most of the applications we envision.

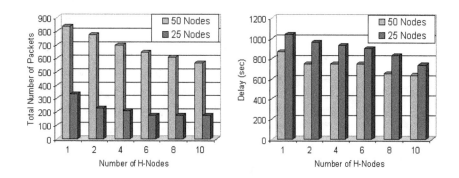

Figure 3.12: The impact of changing the number of H-nodes on (a) the total number of messages, and (b) overall delay.

H-nodes

After analyzing Figure 3.11, we choose the Passive Cure + TTL + BP as our preferred scheme in sparse mobile networks. We use this scheme for our tests on the extended H-node architecture. We show results for sparse networks, setting the number of nodes to 25 and 50. We simulate cases where H-nodes are assumed to represent ultimate nodes. In other words, the goal of the S node is to get the message to any H-node in the sparse network.

Figure 3.12 shows the impact of varying the number of H-nodes in a network of 25 and 50 nodes on our metrics. Both Figures 3.12(a) and 3.12(b) depict a general trend of a decrease in the total number of messages and overall delay as the number of H-nodes is increased. This result is due to the fact that, with a larger number of H-nodes, the message has a better chance of reaching an AP, and

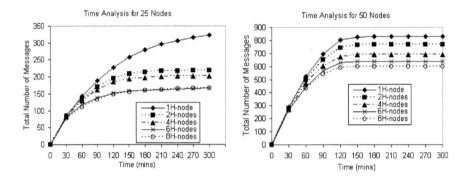

Figure 3.13: Time series analysis of the total number of messages when using H-nodes, with (a) 25 nodes, and (b) 50 nodes.

so, the message can be delivered faster. The total number of messages decreases, because when the number of H-nodes increases, we have more nodes that can "simultaneously" heal the network using the Passive Cure.

More interesting results can be observed in Figure 3.13. The figure generally shows a time series analysis of the total number of messages that have been sent in the network at specific points in time. Figures 3.13(a) and 3.13(b) show the impact of changing the number of H-nodes on the cumulative total number of messages sent at a given time in networks with 25 and 50 nodes, respectively.

Generally speaking, the message spreads faster in denser networks. This result can be seen by the sharper increase in the total number of messages in the 50 node network. Also, the network heals faster in denser networks. This result can be seen by the faster convergence of lines in the 50 node network when compared to that of the 25 node network. This convergence is due to the healing of the network, after reaching a point where all the nodes are healed (or still infected but cannot find anyone to pass the message to), and no more messages are sent through the network. Finally, we observe the intuitive behavior of how increasing the number of H-nodes would generally decrease the total number of messages that are injected into the network. This result occurs due to the increase in the number of high-end nodes that heal the network at a higher rate.

3.4   Summary

In this chapter we have studied the problem of efficient message delivery in delay tolerant sparse mobile networks. We have proposed several controlled flooding schemes on top of a transport layer overlay architecture. The specific schemes we have examined are basic probabilistic, time-to-live, kill time and Passive Cure. We have studied the impact of these schemes on network efficiency and overall message delivery delay. Our simulations have demonstrated that for a given sparse mobile network, the schemes reduce the number of messages and beacons sent in the network. This occurs with either no increase or only a small increase in the overall message delay. We then chose the scheme that performs best, which turned out to be a combination of the Passive Cure with time-to-live and basic probabilistic schemes, and stress tested it. We have examined this combination scheme over an extended architecture that we have proposed, to accommodate for other real-world scenarios.

With the completion of this portion of our work, we believe that we have significantly added to the literature on disconnected and sparse mobile networks, and have opened several new directions of research. Future work in this area includes addressing security issues in such scenarios since such networks are subject to denial of service attacks. Also, since flooding-based schemes do not perform well in dense environments, we intend to develop measures to help nodes modify their behavior when they enter densely populated areas.

Chapter 4

Messenger Scheduling in Clustered Mobile Networks

4.1 Introduction and Motivation

We take the work presented in the previous chapter to the following logical level where we address message delivery problems that exist in other scenarios such as disaster relief efforts and field hospitals [1], battlefields [61], and remote disconnected villages [2], [4]. In these scenarios, we observe a class of Delay Tolerant Mobile Networks (DTMNs) [45] where nodes form clusters such that a communication path exists between any two nodes *within* each cluster. Nodes in *different* clusters, however, cannot communicate with each other except through long-range and high-power wireless or satellite networks. This limitation often occurs because existing infrastructure is either destroyed (e.g., after a hurricane or earthquake), or simply does not exist as in battlefield areas or remote disconnected villages. Also, if the cost of providing these forms of communication is too high or the type of data cannot be transmitted over such networks, these clusters would then need to use other communication paths and methods. This issue becomes particularly important in cases where large amounts of data, such as images of disaster areas or video surveillance clips, must be transmitted between these clusters.

For such scenarios, we envision a new network environment comprised of *regions* and *messengers*. A region is defined as a cluster of nodes having an end-to-end path between any two nodes in the cluster. Since regions are assumed to be disconnected from each other, we propose using a dedicated set of messengers that relay messages between regions. Each region generates large amounts of data that can be grouped together in *bundles* [34] which are then relayed to other destination regions. Multiple messengers would provide fault tolerance and faster delivery of message bundles. The regions could either be mobile, as in search-and-rescue groups or military battalions, or stationary, as in field hospitals or remote disconnected villages. In these environments, we shift the focus from routing [52] and path discovery [110], to *messenger scheduling*. We study how messenger scheduling can be used to improve network performance and connectedness. Furthermore, there are a number of scheduling algorithms (e.g., batching) that can be drawn from marginally related areas and used to provide insight into how best to schedule messengers.

In this chapter, we develop several classes of scheduling algorithms that can be used to achieve inter-regional communication in the environments described above. The novelty of our work is in two main areas. First, we consider mobile environments where regions are mobile and dynamic in nature, as opposed to single stationary or mobile nodes. Second, we introduce the idea of using dedicated messengers under different scheduling schemes rather than discovering and maintaining routing paths in these kinds of networks. We propose a two-dimensional set of solutions representing different scheduling algorithms for messengers. We describe the algorithms in detail, analyze their performance, and derive their bounds and restrictions. Using simulation, we evaluate these algorithms under different network conditions and study their performance in terms of delay, cost, and efficiency. We also examine *adaptive* scheduling strategies to handle sudden behavioral shifts in the system. Our main goals are to gain a better understanding of the challenges involved in the complex network environments we introduce, and to identify which scheduling algorithms are most suitable for different networks.

4.2   System Architecture

Our goal in this section is to set the boundaries for our work and to define the concepts and ideas we introduce. We first briefly describe Delay Tolerant Mobile Networks (DTMNs). We then define and clarify the notion of *regions* and *messengers*. Finally, we discuss different messenger ownership schemes and scheduling strategies and show how these two sets can be merged to create the scheduling strategies we study.

4.2.1   Regions and Messengers

When observing DTMNs, we can see that there are many cases where clusters of nodes, or "hot spots", are formed. These clusters are either mobile or stationary. More importantly, these clusters have end-to-end paths between nodes within the cluster.

Figure 4.1 shows an example with *regions* and *messengers*. A region is an intra-connected cluster of nodes or hot-spot that exists in an extreme environment. A region could represent a military battalion, a remote village, or a disaster response team. Region formation, and the membership of a node in a region, are administrative decisions that are taken depending on the nature of the underlying system. The transport layer protocols used for intra-regional communication is a decision

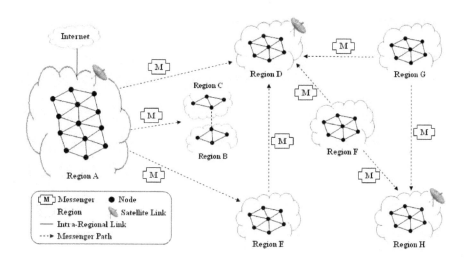

Figure 4.1: Diagram demonstrating regions and messengers in a clustered delay tolerant mobile network (DTMN).

left to the region itself; each region selects what best suits its class of applications and requirements. Since regions are defined to be mobile, they can dynamically divide or merge. For example, in Figure 4.1, Regions B and C are about to merge. For simplicity and without loss of generality, we focus only on cases where regions are mobile, but do not split or merge.

Messengers, also shown in Figure 4.1, are the entities responsible for achieving inter-regional communication. They are dedicated to carrying *bundles* of data across regions. A bundle is a collection of messages that needs to be sent from one or more nodes in a given region to nodes in another region. These messengers could be helicopters, drones, robots, motorbikes, busses, trains, or even planes depending on the application and environment.

A key requirement of our scheduling algorithms is that messengers know the location of other regions and their movement. To begin, each region can receive information regarding its location through the Global Positioning System (GPS). We assume that at least one node in each region will have GPS, which is sufficient for a region to know its own location. The current location of a region could then be transmitted to the messengers through various long-range communication technologies. This requirement is not unreasonable for any of the scenarios or environments we consider. Military battalions or disaster recovery teams, for example, will have at least some basic communication facility for emergency communication. Our contention, however, is that this facility is insufficient for large file transmission or secure communication, but is certainly sufficient for transmitting lightweight data such as region location coordinates.

At this point, we note that we are not addressing routing issues between regions. Routing has been studied in DTN environments [52]. Path determination for ferries in DTNs has also been examined [110]. Routing in the environments we envision could be solved by reducing the image in Figure 4.1 to a graph with each region represented by a single node and then finding the shortest paths between each pair of nodes. Edge weights could determine the cost of a given path between

68

two regions. The cost, for example, could be a function of the distance, difficulty of traversing the terrain, or cost of operating the messenger. These issues are dealt with in other work and are outside the scope of our work in this chapter. We focus instead on the question of when to schedule the movement of messengers between regions.

Finally, we choose to look at such DTMN environments in a distributed fashion, where messengers are scheduled independently in each region. We then propose and study different assignment and scheduling strategies for messengers. Merging these two sets of strategies leads to scheduling strategies that operate dynamically and can be adapted to varying network demands.

4.2.2 Messenger Ownership Schemes

In this section, we present the different messenger ownership schemes and scheduling strategies we propose for achieving inter-regional communication in DTMNs. We propose six different strategies by which messengers are assigned and scheduled. These approaches are summarized by the grid shown in Table I. The Y-axis in the grid represents the ownership of the messenger, while the X-axis describes the time basis upon which a messenger will move from a source region to its destination. The grid can be expanded to include other possibilities

Table 4.1: Ownership and Scheduling Strategies

Scheduling Time / Ownership	Periodic	On-Demand	Storage-Based
Regional	RP – Regional Periodic	ROD – Regional On-Demand	RSB – Regional Storage-Based
Independent	IP – Independent Periodic	IOD – Independent On-Demand	ISB – Independent Storage-based

resulting in a larger spectrum of solutions. We find, however, that the approaches presented in Table 4.1 represent the major points within this spectrum.

The Y-axis in Table 4.1 describes how messengers in a DTMN system are assigned to different regions in the network. We present two alternative assignment strategies: regional messengers and independent messengers.

Regional Messengers mean that each messenger is *owned* by a certain region in the system. The messenger belongs to a given region which is the source region of a given bundle of messages. This messenger can carry these bundles from the source region to a destination region. However, the regional messenger can carry messages from a destination region only back to the source (home) region. Once a messenger reaches a destination region, it delivers the message bundle and then immediately returns to its home region. Regional messengers can, therefore, only carry messages that are either sent by its owner, or destined to its home region from the destination region to which it was initially sent.

Independent Messengers mean that messengers are not owned by any one region. The messenger is managed and directed by the region where it currently resides. We consider this strategy to be a form of temporary ownership by a given source region. Ownership ends once the source region sends a bundle to a new destination. The ownership then transfers to the destination region. Once a messenger reaches a destination region, it delivers the message bundle and then resides in this new region. Independent messengers, therefore, can basically carry messages from one region to any other region in the network. Regions, on the other hand, do not own any messengers, but simply use the available set of messengers to deliver data bundles that need to be transmitted.

4.2.3   Messenger Scheduling Time

The X-axis in Table 4.1 describes the different scheduling strategies with which a messenger could be assigned. We present three alternative scheduling strategies: *periodic*, *on-demand*, and *storage-based* scheduling.

Periodic time scheduling is equivalent to a shuttle system, where shuttles depart according to either a fixed time schedule, or after a fixed amount of time. We use the periodic time schedule to indicate a fixed amount of time after which messengers in a given region are set for departure to a certain destination region. The messengers are only sent if they have message bundles that need to be transmitted, otherwise, they wait for the next period.

On-Demand time scheduling means that messengers are sent to another region as soon as the source region has any message to send. While in the other approaches, messages from multiple sources within the same region are bundled together to be delivered to a given destination, in this case, however, the messenger, assuming one is available, travels as soon as any node within the source region has a message that is required to be sent. This style of system can be viewed as a high-priority system.

Storage-Based time scheduling falls somewhere between periodic and on-demand scheduling. A messenger in this case starts to move towards its destination as soon as it has a predefined amount of data to be delivered. Tweaking the data size requirement, or what we refer to as the "storage limit", affects overall system performance and efficiency.

4.2.4   Merging Strategies and Operation Algorithms

Merging the selections of each type of messenger ownership and scheduling algorithm results in six different approaches as shown by the grid in Table I. Generally speaking, the characteristics of each approach is a result of merging the properties of the two corresponding ownership and scheduling strategies. The

```
1: // Notation
2: Region = R, Message = m, Messenger = M,
3: Messenger Queue = Q, Messenger Ownership = M_O,
4: Max # of regions = n

5: // Message generation and allocation in each region
6: for every R_i where i goes from 1 to n do
7:    for every m_{R_i→R_d} destined to R_d, d ∈ {1..n}, d ≠ i do
8:        Find-Messenger:
9:        Search Q_{R_i} for a messenger M_{R_d} going to R_d;
10:       if (M_{R_d} not found) then
11:           Find M_{R_Null}; // Messenger with no destination set
12:           if (M_{R_Null} found) then M_{R_Null} ← M_{R_d};
13:       if (Q_{R_i} is empty) || (M_{R_Null} not found) then
14:           Wait for Messengers to arrive to R_i;
15:           GOTO Find-Messenger;
16:       // At this point a proper messenger M_{R_d} is found.
17:       Load m_{R_d} → M_{R_d};
18:       if (Ownership == Regional) && (M_O == Null) then
19:           M_O ← R_i;
```

Figure 4.2: Algorithm for loading messages generated in each region onto an appropriate messenger.

high-level algorithms that illustrate the operation of regions, message assignment, messenger dispatch, and messenger travel are shown in the pseudo-code given in Figures 4.2, 4.3, and 4.4.

Figure 4.2 shows the algorithm for loading messages that are generated from various nodes within a region onto a messenger in the source region. In Lines 1-4, we introduce notation that we use in our algorithms. Next, Lines 5-19 show the overall algorithm, which is applied to every message generated in the network.

For each message sent to a remote destination region, an appropriate messenger needs to be discovered (Lines 6-8). We then search in the source region's queue of messengers (Line 9). If no such messenger is found, we search the queue again for a messenger that is not destined to any region yet, and if found, set the destination for this messenger to that of the generated message (Lines 10-12). At this point, if no empty messenger is found, and the messenger queue in the region is empty, there are currently no messengers in the source region that could send the message. In such a case, we wait for messengers to arrive. Then, we perform the search operation again (Lines 13-15). Finally, when an appropriate messenger is found, we load the message and set the ownership of the messenger to that of the source region (Lines 16-19). At this point, the messenger is ready to be dispatched.

Figure 4.3 summarizes the messenger operation before it is sent to a destination region based on the scheduling algorithm of the system. In Lines 20-22, we add new notation to that shown in Figure 4.2. Next, Lines 23-39 show the algorithm in which each messenger in the system is loaded with a given number of messages (previously shown in Figure 4.2), and that these messengers are ready to be sent to their destinations. Based on the scheduling algorithm for the system (Line 26), messengers are appropriately dispatched. If the strategy is periodic, we set the timer for a given period of time. Within this period, the messenger accepts new messages from the region it resides in, so long as its maximum capacity, which we assume to be sufficiently large, is not exceeded (Lines 28-30). After the period is over, the messenger is dispatched to the destination region (Lines 31-32). On the other hand, if the strategy is storage-based or on-demand, the storage limit of each messenger is set accordingly (Line 35). Messages are loaded so long as the capacity is not exceeded. The messenger is sent once the storage limit is met (Lines 36-38). At this point the messengers are dispatched to their intended destination regions.

Figure 4.4 illustrates the messenger operation after it has been dispatched and sent to a destination region. In Lines 40-42, we add new notation to that shown in Figures 3 and 4. Next, Lines 43-58 show the overall procedure that each messenger follows after it has been dispatched. In general, so long as the messenger's location is not equal to that of the destination region, the messenger travels to the last known location for the destination region (Lines 44-45). Upon arrival, the messenger's location is updated (Line 46), and messages are delivered (Lines 47-50). Otherwise, the destination region has moved, and the messenger needs to travel to the new location (Line 44). Upon reaching the final destination, and based on the ownership scheme, the messenger will either stay at the destination region, or return to its owner (home) region. If the ownership is "regional", the messenger gathers all messages in the destination region that need to be sent to the messenger's home region, returns to its home region, delivers the messages,

20: // Notation (added to those in the previous Algorithm)
21: Max # of messengers $= k$, Period $= T$,
22: Capacity of messenger $i = M_{i_C} \leftarrow \infty$

23: // Messenger action based on scheduling algorithm
24: **for** every M_i where i goes from 1 to k,
25: where $n \leq k \leq 2n^2$ **do**
26:　　**switch** (strategy) {
27:　　　**case** Periodic:
28:　　　　Set period timer τ to signal after time T;
29:　　　　**while** $(\sum m \; \epsilon \; M_i < M_{i_C})$ && $(\tau \neq 0)$ **do**
30:　　　　　accept m;
31:　　　　**if** $(\sum m \; \epsilon \; M_i > 0)$ && $(\tau == 0)$ **then**
32:　　　　　Send $M_i \rightarrow R_d$;
33:　　　　**break**;
34:　　　**case** Storage-Based || On-Demand:
35:　　　　Set storage limit M_{i_C} to β; // if on-demand, $\beta \leftarrow 1$
36:　　　　**while** $(\sum m \; \epsilon \; M_i < \beta)$ **do** accept m;
37:　　　　Send $M_i \rightarrow R_d$; // The storage limit is reached.
38:　　　　**break**;
39:　　}

Figure 4.3: Messenger operation, based on scheduling algorithm, before being sent to a destination region.

and joins the messenger queue in the home region to prepare for the next dispatch (Lines 51-56). Otherwise, if the ownership scheme is "independent", the messenger simply joins the messenger queue of the region in which it currently resides.

The six strategies shown in Table 4.1 are straightforward and efficient for the set of applications that we study. Algorithm simplicity is a necessity due to the

40: // Notation (added to those in the previous Algorithm)
41: Messenger Location = L_M, Region Location = L_R,
42: Source Region = R_s, Destination Region = R_d

43: **procedure** Send (M_i, R_d) {
44:    **while** $(L_{M_i} \neq L_{R_d})$ **do**
45:       M_i travels to L_{R_d};
46:       Update L_{M_i} and L_{R_d};
47:       **if** $(L_{M_i} == L_{R_d})$ **then**
48:          Deliver messages $m \in M_i$ to nodes in R_d;
49:          Empty Messenger M_i;
50:          break;
51:    **if** (Ownership == Regional) **then**
52:       Collect messages in current R destined to M_O;
53:       Swap R_s with R_d of M_i;
54:       Return M_i to R_s;
55:       Deliver messages $m \in M_i$ and empty M_i;
56:       Join Q_{R_d} and Set $M_i \leftarrow M_{R_{Null}}$;
57:    **else** Join current region queue Q_{R_s};
58: }

Figure 4.4: Messenger action when sent to a destination region.

complexity of the underlying system in general, as well as the scarce amount of information we assume to be available. Our algorithms are efficient when the proper ownership scheme and scheduling algorithm are matched with the traffic patterns and operational characteristics of the system. The main challenge, therefore, lies in choosing, and then optimizing the most appropriate strategy for a given application.

4.2.5 Adaptive Scheduling

Until this point, we have assumed that the system is configured with a prede-termined ownership and scheduling algorithm. Therefore, the resulting algorithm is not adaptive. Its operation does not change. We make this assumption for two reasons. First, we believe that the most suitable algorithm will initially be chosen for the given environment in which it runs. Second, before we can study adaptive algorithms, we must first understand each of the static algorithms.

Nevertheless, there are many applications that normally run with an initial configuration but are expected to behave in a totally different manner under certain emergency conditions. We assume that during such emergency conditions, at least some parameters will vary outside their nominal ranges. For example, the message rate of the system might increase dramatically for a certain period of time.

There are many ways with which the system can adapt to such emergency activity, however, we only present a simple adaptive strategy that is bounded by the chosen scheduling algorithm of the system. The reason for this bound is because we believe that a total shift in strategy within the system will take significant time to stabilize, which would result in slow adaptation. Therefore, we limit the scheduling adaptation to dynamically changing the storage limit in the storage-based algorithm, or the periodic time in the periodic algorithms. By

78

changing these values as a result of an emergency condition, the messengers adapt to any sudden changes and perform more efficiently. The on-demand algorithm is generally self-adaptive since it already responds to the message generation rate.

Messengers can identify parameter changes by either detecting a sudden change in message generation rate beyond a preset threshold, or receiving a special emergency message through its back haul link. In a similar way, messengers can detect the end of an emergency phase. We show in the evaluation section that this small change in the scheduling algorithm parameter, produces a significant improvement in performance when compared to the non-adaptive strategy.

4.3   Simulation Setup and Environment

The primary goal of our evaluation is to measure the performance tradeoffs of our scheduling strategies. We also identify which scheduling and ownership schemes are more appropriate for different network conditions and environments. We first describe our simulation setup and environment. We then present a deterministic and stochastic analysis for our metrics. Finally, we discuss and analyze our simulation results.

In our simulation setup, we have n regions randomly distributed over a $10km$ x $10km$ terrain. This area is realistic for a rescue team or military battalion, for example. We also have k messengers that are initially equally distributed over the regions in the system. The regions are mobile, and use a *modified* random waypoint mobility model. In our modification, we particularly avoid the major problem of the network slowing down in the conventional random waypoint model. Also, while random waypoint may not be the best model for individual mobile nodes, we believe that it suits the regional mobility patterns with which we are concerned (e.g., for disaster relief). Each region has a diameter, r, that determines the size, and transmission range of the region. Finally, we assume that at least one node in each region has GPS, and therefore, messengers, also equipped with GPS, can dynamically track the location of all regions.

The parameters we believe have the most impact on our system are summarized in Table 4.3. Generally speaking, the speed of messengers is larger than that of regions. We have observed that the actual speed range for regions and messengers does not have much impact on the *relative* performance of the scheduling strategies. Since we are only interested in the relative performance and tradeoffs between these strategies rather than their absolute performance, we do not show the impact of changing these speed ranges. We study the impact of different ownership and scheduling strategies on our metrics. We also consider uniform and exponential message generation patterns for each region.

Table 4.2: Messenger Scheduling Simulation Parameters.

Parameter	Value Range	Nominal Value
Terrain	$10km$ X $10km$	$10km$ X $10km$
Simulation Time	1 hour to 24 hours	12 hours
No. of Regions (R)	3 to 10	5
No. of Messengers	R to R^3	R^2
Message Rate	30 to 150 msgs/h	100 msgs/h
Message Pattern	Uniform, Exponential	Exponential
Region Speed Range	$4km/h$ to $40km/h$	N/A
Messenger Speed Range	$50km/h$ to $90km/h$	N/A
Traffic Pattern	Distributed Uniform or Many-to-one	Distributed Uniform
Scheduling Strategy	Periodic, SB, OD	N/A
Ownership	Regional, Independent	N/A
Periodic Time	5 mins to 120 mins	30 mins
Storage Limit	10 to 1000 messages	50 messages
Emergency Duration	15min to 3hours	1 hour
Emergency Message Rate	500 to 1000 msgs/h	1000 msgs/h

We perform experiments over different inter-regional traffic patterns. We particularly study *many-to-one* traffic and *distributed-uniform* traffic patterns. In the many-to-one pattern, many regions send messages to one region in the system, while in the distributed-uniform pattern, all regions have an equal probability of sending messages to any other region. In this way, we satisfy scenarios where several regions need to send data to a central region (e.g., a battlefield command center) or all regions simply distribute their data to all other regions (e.g., search and rescue teams). We also vary the message generation rate in our system to observe the performance under different load levels.

Many applications targeted by our system experience various extreme conditions. These conditions result in sudden changes in the behavior of a system. We call this sudden behavioral change an emergency. Such emergencies are represented by the *emergency message rate*, which is a temporary high message rate at which the system operates for an *emergency duration* amount of time.

We consider three metrics in evaluating the scheduling algorithm tradeoffs in our system. These metrics are:

- **Delay:** measured by the average time taken by a message to go from its source region to a destination region. This includes the wait time of a message at its regional queue, as well as the travel time on the messenger.

- **Cost:** measured by the total distance or the total number of trips the messengers have taken after a given amount of time. We use distance in our analysis and number of trips in our simulation results. While we also measured the total distance in our simulations, we only show the number of trips since they have shown similar relative behavior with respect to our algorithms.

- **Efficiency:** measured by the average number of messages carried and delivered in each trip. This metric measures messenger efficiency.

4.4 System Analysis

In this section we theoretically analyze our algorithms and metrics for evaluation. The ultimate goal of this section is to present formalized and probabilistic bounds on our metrics. To accomplish this goal, we first state some simplifying assumptions that help us present our results supposing determinist conditions. We then use these derived results to conduct stochastic analysis leading to a set of interesting probabilistic estimates and bounds.

4.4.1 Deterministic Analysis

In this section we deterministically analyze the three metrics of delay, cost, and efficiency in our system. We perform this analysis for each of our algorithms.

Delay

The delay of a message, based on our definition, is:

$$\text{Delay} = t_{wait} + t_{travel} + 2 \times t_{load}. \tag{4.1}$$

In the equation, t_{wait} is the time a message waits at its source region. This time includes the queuing time at the regional queue plus the time a message may spend on a messenger waiting for the period to expire, or for the storage limit to be reached. Also, t_{travel} is the time it takes a messenger to deliver a message. Finally, the loading and unloading time, is denoted by t_{load}. This loading time is negligible when compared to the other two values. We suppose that messengers going from R_i to R_j travel at a constant speed, v, until they reach their destination.

In all the algorithms, the travel time between a source, R_i, and a destination, R_j, will be determined by $t_{travel} = d_{ij}(k)/v$, where d_{ij} is the distance between the two regions at time t_k. We now compute t_{wait} for each algorithm and Equation (4.1) can then be used to obtain the total delay.

On Demand: The upper-bound for the wait time for messages at R_i that need to be delivered to R_j using the regional ownership is roughly:

$$t_{wait} = \frac{d_{ij}(k-1) + d_{ji}(k)}{v}. \tag{4.2}$$

Since messages can sometimes wait for the return of the messenger from R_j, we consider both distances in the equation. In such cases, the distances can change between two different legs of travel, so $d_{ij}(k-1)$ is the distance between R_i and R_j at time t_{k-1}, and $d_{ji}(k)$ is the distance at time t_k. The wait time at R_i for the messages using independent ownership is:

$$t_{wait} = \frac{d_{qi}}{v}. \tag{4.3}$$

In this equation, the distance d_{qi} is from any region R_q whose messenger arrives at R_i in time to collect queued messages destined to R_j.

Periodic: In this part we express the wait time for both the regional and independent ownership algorithms in the following equation:

$$\sup_{d_k} t_{wait} = \max(T, \frac{\alpha d_k}{v} + \frac{d_{k-1}}{v}), \ where \ \alpha \in (0, 1). \tag{4.4}$$

In this equation, α is either 0 or 1 for the independent and regional ownerships strategies, respectively, since we consider one messenger trip for the independent and both trips for the regional. The sup is the supremum function, that is, the smallest upper bound between the period, T, and the t_{wait}. The wait time is introduced this way so that if the messenger return time is larger than the period, we do not falsely assume the messenger to be available.

Storage-Based: Similar to the periodic algorithm, we calculate the supremum message wait time for the storage-based algorithm with the maximum between the time to reach the storage limit and the messenger travel time as:

$$\sup_{d_k} t_{wait} = \max(t_{storage}, \frac{\alpha d_k}{v} + \frac{d_{k-1}}{v}), \ where \ \alpha \in (0, 1). \tag{4.5}$$

We now need to calculate $t_{storage}$, the time it takes to reach the storage limit of a messenger. We suppose s_{limit} is the maximum storage per messenger. We now compute the time it takes to generate s_{limit} messages, given a message generation rate of λ_i at region i, where $i \in \{1, 2..n\}$. To calculate $t_{storage}$, we solve for t, given $\lambda_i(t)$:

$$s_{limit} = \int_{t_0}^{t_{limit_i}} \lambda_i(t) \ dt = \Lambda_i(t_{limit_i}) - \Lambda_i(t_0) \qquad (4.6)$$

$$\Rightarrow \Lambda_i(t_{limit_i}) = s_{limit} + \Lambda_i(t_0), \ where \ \Lambda_i(t) = \int \lambda_i(t). \qquad (4.7)$$

If we suppose λ_i to be a constant, at least in the interval of integration, the integral can be rewritten as:

$$s_{limit} = \int_{t_0}^{t_{limit_i}} \lambda_i \ dt = \lambda_i(t_{limit_i} - t_0)$$

$$\Rightarrow t_{storage} = t_{limit_i} - t_0 = \frac{s_{limit}}{\lambda_i}. \qquad (4.8)$$

In these equations, the storage time, $t_{storage}$, is the waiting time until the number of messages reaches s_{limit} in a messenger at region i, at a time t_{limit_i} starting from not having any messages at t_0.

Cost

The cost of the algorithm is the overall distance traveled by all the messengers. It is given by the following equation:

$$\text{Cost} = \sum_{i=1}^{m} \int_{0}^{t_{final}} f(t) \; v_i(t) \; dt \tag{4.9}$$

$$where \; f(t) = 1 \quad for \; t \in [t_{(k-1)}, t_{(k)}], \; 0 \; otherwise.$$

In this equation, $v_i(t)$ is the velocity with which messenger i travels during the time interval $[t_{(k-1)}, t_{(k)}]$, where t_k take values from $[0, t_{final}]$. t_{final} is the end time after which the system halts, or is put to a stop. During this interval, we calculate the distance traveled by a messenger i. We then integrate over the whole time of operation to obtain the total distance covered by this messenger. Finally, we use the outer summation to obtain the total cost incurred by all m messengers.

Efficiency

The efficiency of the system is the number of messages carried per unit distance. Assuming a message generation rate of λ_i at each region i, the number of messages, μ, generated at this region is:

$$\mu_i = \int_0^{t_{final}} \lambda_i(t) \, dt. \tag{4.10}$$

The overall efficiency of the system is then the total number of messages generated and delivered by all the regions in the system, divided by the total distances
covered by all the messengers. The equation for efficiency would then be:

$$\text{Efficiency} \quad = \quad \frac{\sum_{i=1}^n \int_0^{t_{final}} \lambda_i(t) \, dt}{\sum_{i=1}^m \int_0^{t_{final}} f(t) \, v_i(t) \, dt}. \tag{4.11}$$

The numerator and denominator are taken from Equations 4.10 and 4.9, respectively. f(t) is defined the same way as in Equation 4.9. We sum all the
messages generated over all the regions and let the time go from $[0, t_{final}]$. The
terms in the denominator are the same as in Equation 4.9, and they account for
the overall distance traveled during the whole time interval for all messengers.

4.4.2 Stochastic Analysis

So far, we have assumed the system parameters to be functions of time, and we
have considered them to be deterministic functions of time. In order to conduct
our stochastic analysis, we first present the probability distributions of our random

variables. We then use these distributions as well as our deterministic analysis as

a base for our stochastic analysis.

Probability Distributions

We focus on the two variables of our probability distribution that have the

most impact on our metric calculation, namely, the message generation rate, and

the distance between regions. We relax the assumption of deterministically in-

corporating these parameters, and assume these variables to be random variables

with given probability distributions.

Message generation rates: The deterministic generation rate $\lambda_i(t)$, of messages

is now assumed to have a poisson distribution with an average number of messages

per unit time. We then know the probability of having c messages generated before

a given time, t, to be:

$$P(c) = \frac{e^{-\lambda t}(\lambda t)^c}{c!}, \quad where\ c \in (0, 1, 2, \cdots).\qquad(4.12)$$

Distance between regions: Instead of assuming random velocities, we assume

random distances between the regions. Since we are using the random way point

mobility model, we build on the analysis presented by Bettstetter et al. [13], and

obtain the probability distribution of the distances between regions in a $10km$ x

$10km$ terrain to be as shown in the following Figure 4.5:

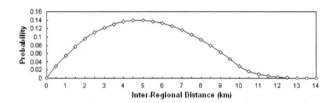

Figure 4.5: Probability distribution for inter-regional distances when using the random way point mobility model.

Based on Bettstetter et al. [13], it is equally likely for each region to start at any point in the terrain. Also, to obtain the coordinates of each region, the coordinates, P_x, of any point in the square have the following Probability Density Function (PDF), where a is the squared terrain dimension. $f_{P_x}(X) = \frac{1}{a}$ for $0 \leq x \leq a$, 0 otherwise. The joint distribution of two independent points is then given by $f_{P_{x_1} P_{x_2}}(X) = \frac{1}{a^2}$ for $0 \leq x_1, x_2 \leq a$, 0 otherwise. The distance, D, between two regions at positions $P_{(x_1,y_1)}$ and $P_{(x_2,y_2)}$ is given by $d = \sqrt{L_x^2 + L_y^2}$.

The joint probability distribution of the distance between two regions is given by:

$$f_{L_x L_y}(l_x, l_y) = f_L(l_x) f_L(l_y) = \frac{4}{a^4}(-l_x + a)(-l_y + a). \qquad (4.13)$$

According to Bettstetter et al. [13], the Cumulative Density Function (CDF) of the distances between two regions will then be given by: $P(d \leq l) = \iint_D f_{L_x L_y} \, dl_y \, dl_x$. So that we have:

$$P(d \leq l) = \begin{cases} \int_0^l \int_0^{\sqrt{l^2 - l_x^2}} f(l_x)(l_y) \, dl_y \, dl_x. & 0 \leq l \leq a \\[2em] \int_0^a \int_0^{\sqrt{l^2 - a^2}} f(l_x)(l_y) \, dl_y \, dl_x \\ \quad + \int_{\sqrt{l^2 - a^2}}^a \int_0^{\sqrt{l^2 - l_x^2}} f(l_x)(l_y) \, dl_y \, dl_x. & a \leq l \leq \sqrt{2a^2} \end{cases} \tag{4.14}$$

Taking the derivative of the CDF, we obtain the PDF of the distance:

$$P(d = l) = \begin{cases} \frac{4l}{a^4} \left(\frac{\pi}{2} a^2 - 2al + \frac{1}{l^2} \right) & 0 \leq l \leq a \\[1em] a^2 [\arcsin[\frac{a}{l}] - \arccos[\frac{a}{l}]] + 2a\sqrt{l^2 - a^2} - a^2 - \frac{l^2}{2} & a \leq l \leq \sqrt{2a^2} \end{cases}$$

Finally, the expected value of the distance can be obtained by computing $\int_0^{a\sqrt{2}} P(d = l) dl$. Solving this would result in the following equation:

$$E[d] = 0.5214a \tag{4.15}$$

Performance Computations

Given all of the expressions and distributions we have presented for our metrics and variables, we now analyze and present the probabilities of some interesting indicators of system performance.

- Delay. We calculate the probability of the wait time being smaller than or equal to a given value, β, for each of our algorithms. We note that for the following three equations, we use $\alpha \in (1, 2)$, such that $\alpha = 1$ if the ownership is independent, and $\alpha = 2$ if the ownership is regional. Also, $E[v]$ is the expected value of the messenger velocity.

1. For the **on-demand** algorithm, using Equations (4.3) and (4.14), we obtain:

$$P(t_{wait} \leq \beta) = P(\frac{\alpha d_i}{v} \leq \beta) = P(d_i \leq \frac{\beta v}{\alpha}) = P(d_i \leq \frac{\beta E[v]}{\alpha}) \qquad (4.16)$$

2. For the **periodic** algorithm, using Equations (4.4) and (4.14), we obtain:

$$P(t_{wait} \leq \beta) = P(\alpha \frac{d}{v} \leq \beta) = P(d \leq \frac{\beta v}{\alpha}) = P(d \leq \frac{\beta E[v]}{\alpha}) \qquad (4.17)$$

if the period $T \leq \beta$, 0 otherwise.

3. For the **storage-based** algorithm, using Equations (4.5) and (4.14), we

obtain:

$$
\begin{aligned}
P(t_{wait} \leq \beta) &= P(t_{storage} \leq \beta, \frac{\alpha d}{v} \leq \beta) \\
&= P(t_{storage} \leq \beta | \frac{\alpha d}{v} \leq \beta) \times P(\frac{\alpha d}{v} \leq \beta) \\
&= P(N_{Strorage} = s_{limit}) \times P(d \leq \frac{\beta E[v]}{\alpha}) \\
&= e^{-\lambda \beta} \frac{(\lambda \beta)^{s_{limit}}}{s_{limit}!} \times P(d \leq \frac{\beta E[v]}{\alpha}).
\end{aligned}
\tag{4.18}
$$

In these equations, since the CDF better describes the distributions of random

variables, we choose to compute it. We note that one can use these equations to

obtain the PDF to find $P(t_{wait} = \beta)$ by calculating the difference of two CDF

points. In the storage-based equation, since we are looking for the maximum

of two terms, based on Equation 4.5, to be less than β, we need both terms in

the joint probability function. We expand the joint probability to compute the

joint probability distribution using Bayes conditional probability theorem, which

states that $P(A \cap B) = P(A|B) \times P(B)$. The first term in this expansion is

then equivalent to finding the probability of the number of generated messages

equal to s_{limit}. This expansion is because a messenger must become available

before β, hence, the condition $T \leq \beta$ in the conditional probability. Along with the definition of $t_{storage}$, we obtain the $N_{storage}$ probability. We then solve the probability function using Equation 4.12.

- Cost. In order to determine the probability of the overall cost being less than a certain value, we first need to change the bound on the cost to a bound on the distance traveled per messenger per trip. We use this reduction because we know the probability distribution of the distances between any two regions. Given the average waiting time, t_{wait}, of each each message going from a region, R_i, to a second region, R_j, we estimate the number of times messages going from R_1 to R_2 will be sent in a total simulation or operation time, t_{total}, to be:

$$E[n_{trips}] = \frac{t_{total}}{t_{wait}}. \tag{4.19}$$

Each of these transmissions corresponds to a trip by a messenger. The expected length of the travel is given by Equation 4.15, which gives us an estimate on the distance traveled by messengers taking messages from R_1 to R_2. The expected value of the total cost will then be $E[Cost] = 0.5214an(n-1)$.

In the computation above we were able to derive the expected value of the total cost. We are now interested in calculating the probability distribution of the cost. Since we know that each messenger will on average travel n_{trips}, computed in Equation 4.19, the probability of the total cost being less than or equal to C is given by:

$$
\begin{aligned}
P(Cost \leq C) &= P(E[n_{trips}]n(n-1)d \leq C) \\
&= P(d \leq \frac{C}{E[n_{trips}]n(n-1)}).
\end{aligned}
\tag{4.20}
$$

We note that we can take into consideration the possibility that no messages are generated and no trips will happen within t_{wait}. This probability is given by $e^{-\lambda t_{wait}}$ (we set c to zero in Equation 4.12). So the expected number of trips that will occur becomes, $E[n_{trips}] \times (1 - e^{-\lambda t_{wait}})$. We then take into account the different regions and destinations to compute the total number of trips as $n_{total} = E[n_{trips}]n(n-1)$. This equation can be used to estimate the total cost in distance by using Equation 4.15. The total distance will be $d_{total} = (0.5124a)n_{total}$.

- Efficiency. For the efficiency metric, we compute the efficiency of a single trip by one messenger and obtain the probability of it being greater than a specified value μ. This probability is assumed to approximate the probability that the total efficiency is greater than μ. If the efficiency of every trip is greater than μ, the efficiency of all the trips will be greater than μ, as the distances as well as the messages will be multiplied by the same number $m \times n_{trips}$, to account for all the trips by all the messengers. Below, we compute the probability of the efficiency being greater than μ for a given trip, which is equal to the probability that the number of messages, N_t, is greater than μ_1, and the distance less than μ_2, given that $\frac{\mu_1}{\mu_2} = \mu$:

$$P(Efficiency \geq \mu) \quad = \quad P(N_t \geq \mu_1, d \leq \mu_2) \tag{4.21}$$

$$where \quad \mu_1 = \inf_{x}(x \in \mathbf{Z}),$$

$$and \quad \mu_2 = \sup_{y}(y \in [d_{min}, d_{max}]), \; s.t. \; \frac{\mu_1}{\mu_2} = \mu.$$

Where inf is the highest lower bound, and Z is the set of integers. The probability of having an efficiency $Efficiency \geq \mu$ is then given by:

$$P(N \geq \mu_1, d \leq \mu_2) \quad = \quad P(N \geq \mu_1 | d \leq \mu_2) \times P(d \leq \mu_2)$$

$$= \quad (1 - P(N < \mu_1 | d \leq \mu_2)) \times P(d \leq \mu_2)$$

$$= \quad (1 - P(N < \mu_1 | t \leq \frac{\alpha\mu_2}{v})) \times P(d \leq \mu_2)$$

$$= \quad (1 - \sum_{k=0}^{(\mu_1-1)} \frac{e^{-\lambda(\frac{\alpha\mu_2}{v})}}{k!}(\frac{\lambda\alpha\mu_2}{v})^k) \times P(d \leq \mu_2). \tag{4.22}$$

In the above equation, we again use the multiplication rule to compute the joint probability of the intersection of two events. The expansion of the first term in the first equality is because the set, $S_1 = n \in \mathbf{Z}$, $s.t.$ $N_t \geq \mu$, is the complement of $S_2 = n \in \mathbf{Z}$, $s.t.$ $N_t < \mu$, and the measure of these two sets sums to one, and their intersection is the empty set. Therefore the probability of the occurrence of an event in one of the two sets satisfies $P(x \in S_1) = 1 - P(x \in S_2)$. The rest follows from analysis analogous to the previous sections.

We can provide further analysis and solve for an optimal number of messengers to be used given any of the algorithms. This analysis, however, is beyond the scope of our work, and is more relevant to control theory. In fact, Waisanen et al. work on similar issues by providing lower bounds on delays achievable by controlled policies in networks with controlled mobility [98]. We next describe our simulations and results that support this analysis. After we present our simulation results, we show a representative set of analysis-based results that verify our simulation results.

4.5   Simulation Results

In our evaluation, we conducted an extensive set of simulations, but will focus in this section on the subset that best illustrates the results that lead to our conclusions. We show the tradeoffs between the performance of the different scheduling strategies and ownership schemes that we propose. We first discuss how the performance of the periodic and storage-based strategies vary according to the periodic time and storage limit. Next, we demonstrate the impact of varying both the message generation rate and the number of regions. We look at this impact under the two different ownership schemes. Then, we show the operation of our scheduling schemes under different network traffic patterns. Finally, we investigate how adaptive strategies perform compared to static ones during emergency situations. All measurements are taken with respect to the three metrics of delay, cost, and efficiency. Nominal values of the parameters in Table II are used for all experiments, except for those parameters being tested. Each point in our results is taken as an average of 20 different simulation seeds.

4.5.1 Periodic and Storage-Based Performance

Before showing the relative performance of our scheduling schemes under different network conditions, we first need to describe the performance of the periodic and storage-based schemes. The reason for this choice is because the performance of those two schemes relies heavily on the choice of the periodic time and storage limit values, respectively. This result is evident in both Figures 4.6 and 4.7.

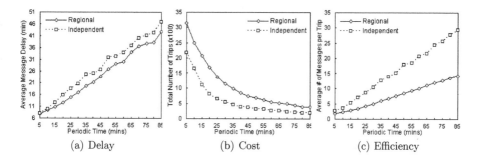

(a) Delay   (b) Cost   (c) Efficiency

Figure 4.6: Impact of changing the periodic time when adopting the periodic scheduling algorithm for both ownership schemes.

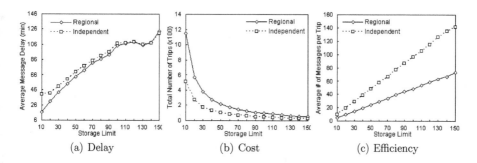

(a) Delay   (b) Cost   (c) Efficiency

Figure 4.7: Impact of changing the storage limit when adopting the storage-based scheduling algorithm for both ownership schemes.

Figure 4.6 shows how varying the periodic time impacts the performance of the periodic scheduling algorithm in terms of our metrics. Figure 4.7 shows how varying the storage limit impacts the performance of the storage-based scheduling algorithm. We measure the performance of these schemes under both regional and independent ownership. In general, Figures 4.6 and 4.7 show that as the periodic time or storage limit increases, respectively, delay linearly increases; cost exponentially decreases; and efficiency linearly increases. An increase in periodic time or storage limit means that messengers travel less often. Therefore, messengers take fewer trips and carry more messages per trip. The results also show that, due to the exponential drop in cost versus the linear increase in delay and efficiency, an optimal period or storage limit exists. For example, a period of 30 or a storage limit of 50 are the most efficient for this setup.

With respect to the impact of the ownership scheme, we observe that changing the ownership scheme does not significantly impact delay. The cost and efficiency, however, vary greatly depending on the ownership scheme used. The small difference in delay is due to the distributed uniform traffic pattern that is used in the system. This pattern causes the regional and independent schemes to operate similarly in terms of delay. Conversely, the regional scheme incurs approximately double the cost and half the efficiency when compared to the independent scheme. The reason for this result is that each messenger in the regional ownership system must immediately return to its owner once it has delivered its set of messages. The messenger then only carries messages that are queued at the destination region and addressed to the messenger's owner region. If the queue at the destination region is empty, the messenger would return with no messages. This result means that messengers, in the worst case, would have to travel approximately double the number of trips compared to independent ownership. We note that while regional ownership may not seem advantageous in these results, it guarantees messenger availability for all regions, where the independent approach can lead to starvation depending on the system's traffic pattern.

4.5.2 Impact of the Message Rate

After demonstrating the performance of the periodic and storage-based scheduling strategies separately, we now compare them to the on-demand strategy while varying the message generation rate. Figure 4.8 shows the impact of changing the message generation rate on our scheduling strategies using the regional ownership scheme. Figure 4.9 shows the same results for the independent ownership scheme.

In general, Figures 4.8 and 4.9 show similar relative behavior in terms of scheduling algorithm performance. Delay in both Figures 4.8(a) and 4.9(a) remains relatively constant for the on-demand and periodic strategies, but decreases as the message rate increases. The reason for this result is that the storage-based strategy is sensitive to the message rate while the others are not. As the message rate increases, the storage limit is reached more quickly resulting in a shorter wait time, and therefore, shorter delays. The other two strategies, however, are insensitive to the change in message rate, because there is a large upper bound on the number of messages that can be carried in these strategies.

With respect to cost, shown in Figures 4.8(b) and 4.9(b), there is an increase in the cost incurred by the storage-based and on-demand schemes. The figures also show that the results for the periodic scheme remain constant. This result occurs because the periodic scheme operates independently of the message generation rate; messengers move after the periodic time expires regardless of the number of messages it carries, and therefore, execute the same number of trips. The on-demand and storage-based schemes are sensitive to the change in message rate, since a higher message rate would cause more on-demand trips, or will reach the storage limit more quickly.

The efficiency of the system, shown in Figures 4.8(c) and 4.9(c), increases for the periodic and on-demand schemes while remaining constant for the storage-based scheme. This result occurs because the efficiency of the storage-based scheme is fixed since the messenger does not move unless its storage limit is reached. However, for the other two schemes, as the message rate increases, more

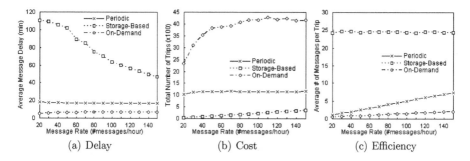

Figure 4.8: Impact of changing the message generation rate on the performance of the three scheduling strategies under the *regional* ownership scheme.

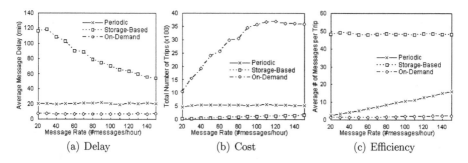

Figure 4.9: Impact of changing the message generation rate on the performance of the three scheduling strategies under the *independent* ownership scheme.

messages can then be loaded when a messenger is waiting (periodic), or more messages can be carried back to the owner region (on-demand).

We observe that under both ownership schemes, the storage-based strategy has the least cost and highest efficiency with the largest delay, while the on-demand strategy has the least delay at the expense of incurring the highest cost and the lowest efficiency. The periodic strategy offers the most balanced performance in terms of our metrics. The choice of a strategy then depends on which metric matters the most to the application. Even though the general behavior of the strategies looks the same under different ownership schemes, the difference in terms of cost and efficiency is large in both cases. This observation can be clearly seen when we compare the Y-axis scales for these two metrics under different ownership schemes. We find that the regional scheme (Figure 4.8) incurs approximately double the cost and half the efficiency when compared to the independent scheme (Figure 4.9), for all scheduling strategies. This result is due to the extra trip each messenger must take to return to its owner region after delivering messages.

4.5.3 Impact of the Number of Regions

We now compare the different scheduling strategies in our system while keeping the message rate constant and changing the number of regions, and consequently, the total number of messengers[1]. Figure 4.10 shows the result of changing the number of regions on our scheduling strategies using the regional ownership scheme. Figure 4.11 shows the same results, but for the independent ownership scheme.

In general, Figures 4.10 and 4.11 show similar relative behavior in terms of scheduling algorithm performance according to our metrics under both regional and independent ownership schemes. However, when focusing on the axis scales, we see that they have different absolute behavior. Delay in both Figures 4.10(a) and 4.11(a) increases for the storage-based scheme while remaining constant for the other two strategies. The reason is again due to its sensitivity to the message rate. As the number of regions increase, while keeping the regional message generation rate constant, there are more destinations to which the same number of messages can be addressed. This relationship means that the messengers will have to wait longer until the storage-limit is reached. The other two strategies remain insensitive as we explained in the previous section.

With respect to cost, shown in Figures 4.10(b) and 4.11(b), we see an increase in the cost for all scheduling strategies. This result occurs because, since we have more messengers to compensate for the increase in the number of regions, there is a resulting increase in the total number of trips. Finally, the efficiency

---

[1]See Table 4.3 for the relationship between the number of messengers and the number of regions.

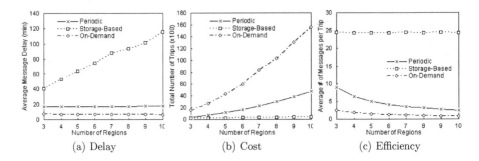

Figure 4.10: Impact of changing the number of regions on the performance of the three scheduling strategies under the *regional* ownership scheme.

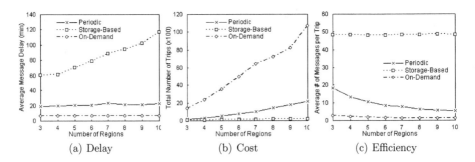

Figure 4.11: Impact of changing the number of regions on the performance of the three scheduling strategies under the *independent* ownership scheme.

of the system, shown in Figures 4.10(c) and 4.11(c), decreases for the periodic and on-demand scheduling strategies while remaining constant for the storage-based strategy. The reasoning is the same as for the previous sets of results. If a messenger moves only when its storage limit is reached, it maintains a constant efficiency level. The efficiency for the on-demand and periodic schemes decreases, however, because the same number of messages are divided over a larger number of destinations. The overall result is a decrease in the number of messages each messenger carries when traveling to a given destination.

Similar to the behavior in the previous section, and for the same reasons, the storage-based strategy has the least cost and the highest efficiency, but with large delay, while the on-demand scheme has the highest cost and the lowest efficiency with the least delay. Again, we observe that the regional scheme (Figure 4.10) incurs approximately double the cost and half the efficiency when compared to the independent scheme (Figure 4.11), for all scheduling strategies.

4.5.4   Impact of the Traffic Patterns

So far, all presented results have been based on the use of a distributed uniform traffic pattern. In this section, we demonstrate the impact of adopting a many-to-one pattern and show results in Figures 4.12 and 4.13.

Figure 4.12 shows the impact of changing the message generation rate on our scheduling strategies using the independent ownership scheme. Figure 4.13 shows the impact of changing the number of regions on our scheduling strategies also under the independent ownership scheme. While the general performance looks similar, the results need to be compared to those in Figures 4.9 and 4.11, respectively, to grasp the impact of using a many-to-one traffic pattern as opposed to a distributed uniform one. In general, there is a reduction in the cost incurred and an increase (except for the storage-based scheme) in efficiency in Figures 4.12 and 4.13 when compared to Figures 4.9 and 4.11, respectively.

The intuition behind this variation in results is as follows. In general, under a many-to-one scheme operating with independent ownership, the environment begins to look like a basic command center, or sink, where most of the messengers gradually accumulate. When these messengers are sent to their destinations, they gather as many messages as possible (except for the storage-based scheme) from those queued at these destination regions. This behavior works well under average to high message generation rates. However, if the message generation rate is low, or the sink does not produce any messages at all, the independent scheme could result in starvation when all the messengers accumulate at the sink and are rarely, if ever, sent to the other regions. The solution in such cases is simple: a periodic timer could be used to re-distribute messengers across the regions in the system.

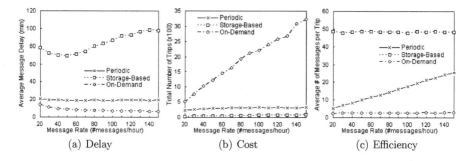

Figure 4.12: Impact of the *message rate* on the performance of the scheduling strategies. Independent ownership and a many-to-one traffic pattern are used.

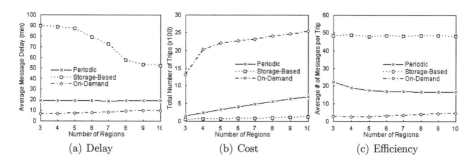

Figure 4.13: Impact of the *number of regions* on the performance of the scheduling strategies. Independent ownership and a many-to-one traffic pattern are used.

109

We do not show results for the regional ownership scheme because the overall performance is similar to that of the distributed uniform traffic pattern. The reason for this result is that each region owns a number of messengers that must return to it upon message delivery.

4.5.5 Adaptive Scheduling Strategies

We now examine the impact of our adaptive scheduling algorithm compared to the static ones we have studied so far. In Figures 4.14 and 4.15, we compare the performance of the adaptive to the non-adaptive algorithms under different emergency durations. Figure 4.14 shows the impact of dynamically changing the storage limit from 50 messages to 100 messages when using the storage based scheduling algorithm. Figure 4.14 shows the impact of changing the periodic time from 30 mins to 15 mins when using the periodic scheduling algorithm. We note that messengers in these adaptive cases change their scheduling parameters only for the duration of the emergency interval, then return to their nominal value of operation. The overall results in these two figures demonstrate a significant improvement in performance when the adaptive strategy is used, particularly in terms of delay, which is usually the main metric we are concerned with during emergency intervals.

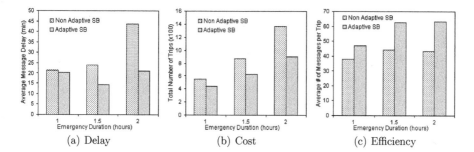

(a) Delay   (b) Cost   (c) Efficiency

Figure 4.14: Comparing the adaptive *regional storage-based* algorithm to the non-adaptive one. The storage limit changes dynamically to adapt to different emergency durations.

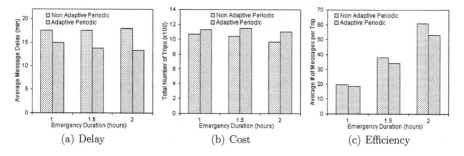

(a) Delay   (b) Cost   (c) Efficiency

Figure 4.15: Comparing the adaptive *regional periodic* algorithm to the non-adaptive one. The periodic time changes dynamically to adapt to different emergency durations.

Both figures show an interesting result: the positive impact of an adaptive strategy is greater when the duration of the emergency time grows larger. The adaptive strategies in both figures satisfy the goal of decreasing delay during emergency times. This result occurs because the larger storage limit or smaller period makes messages wait for less time at the regional queues, which ultimate leads to smaller delay. In fact, the adaptive storage-based strategy does better in all three metrics. We also observe that we incur a small increase in cost in the adaptive periodic strategy simply because the number of trips will increase with the drop in the period duration.

We finally note that the choice of the storage limit or periodic time for the adaptive strategy in these simulations is not optimal. This choice is kept simple by picking a new value based on the amount of change in the system. The reason for this simplicity is because the goal of these simulations is to show that *any* form of adaptation is likely to cause a performance improvement. An optimal choice for the new storage limit or periodic time can be obtained by finding the optimal values for these parameters under different emergency network situations. This observation means that we conduct experiments similar to the first set of results presented in this section for various message rates in the system, and determine the most appropriate storage limit and periodic time values for these rates. Therefore, when an emergency occurs, the system can adapt the optimal parameter values to the new emergency message rate.

4.5.6 Other Results

So far, we have presented a subset of results for cases we believe are common in the applications we consider. However, we conducted numerous other simulations that we do not show in this chapter due to the rareness with which these situations may occur. Extreme conditions with large message bursts or very low message generation rates are good examples of rare cases. Results in such conditions show that on-demand behaves well with very low rates since messengers are sent only when a message is generated. Under large message bursts, the periodic or storage-based schemes seem to operate well. In these cases, adapting within these strategies improves system performance.

Varying the number of messengers while keeping the number of regions fixed is an example of another result analyzed. We only present results where the number of messengers increases with the increase in regions because we believe that the number of messengers deployed will be proportional to the expected number of regions. However, if the number of regions were to increase while maintaining the same number of messengers, delay would increase due to the increase in message queue time. This result occurs because there are more destinations and potentially more messages that need to be served with the same number of messengers.

4.5.7 Verifying Simulation Results

While the simulation results we obtained give a lot of insight with respect to the different tradeoffs that we witness, further confidence can be gained by verifying these results through analysis. We perform this verification by programming and running our analysis formulas in Matlab and comparing the output to those obtained via simulations.

We show a representative subset of our analysis results in Figure 4.16. This subset represents the impact of message generation rate on the regional ownership scheme. The results in Figure 4.16 can be compared to our simulation results depicted in Figure 4.8. We basically observe that plotting the average upper bounds or estimates for our metrics, shown in Figure 14, based on our analysis section, reveal similar behavior to what we obtain in our simulations.

The small differences that may be observed in our analysis results are due to the simplifying assumptions made in our analysis in order to make the analysis tractable. We note that even though our analysis estimates cost in terms of distance, we simply divide this estimate over the expected average length of a trip. Finally, we note that we do not show results for other simulations due

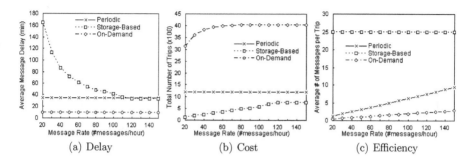

Figure 4.16: Analysis results of the impact of the message generation rate on the performance of the scheduling strategies under the regional ownership scheme.

to space limitation, but have verified the predictability of our simulation results based on our analysis.

4.6   Summary

In this chapter we have studied the idea of using a dedicated set of messengers for message delivery in mobile networks where nodes form disconnected clusters, which we define as regions. Messengers are then used to communicate between these regions. We have introduced two messenger ownership schemes, regional and independent, as well as three scheduling algorithms, periodic, storage-based, and on-demand. Our results from studying these schemes and algorithms have demonstrated that the choice of a particular algorithm ultimately depends on the environment in which it is deployed, as well as which metric is most important to the application being served. While the on-demand algorithm promises the least delay, it comes at the expense of large cost and low efficiency. The periodic algorithm, on the other hand, seems to be a reasonable solution as it maintains the most balanced performance with respect to our metrics. Selecting an appropriate periodic interval, though, is a challenge. The storage-based algorithm generally gives the highest efficiency and least cost but requires setting the storage limit, and comes at the expense of high delay. Finally, we have shown how even the simplest form of adaptive strategies can lead to an improvement in performance.

With the completion of our work in this chapter, we believe that we have opened several new directions of research. Future work in this area includes studying environments where messengers are destroyed or lost, and therefore, require the deployment of end-to-end reliability mechanisms for message delivery. Hence, our algorithms become what we call a first generation of scheduling strategies. More intelligent and adaptive solutions can be designed where messengers can dynamically chose a different scheduling algorithm, or even a mix of strategies, to enable them to adapt to changing network conditions. In the end, as networks are increasingly expected to provide communication in hostile and challenged environments, better and more robust solutions will be needed to fulfill such expectations.

Chapter 5

Transport Layer Issues in Delay Tolerant Networks

5.1 Introduction and Motivation

With the work in DTNs mainly focused on routing, we shift our focus towards studying transport layer issues. Most of the services offered by existing transport layer protocols, such as TCP, have been overlooked in regards to DTNs. In these environments, the most important services offered by TCP are ports, connections, sequencing, congestion control, and reliability. Some of these services are easy to deploy in DTNs, while others require further research.

Of the TCP services previously mentioned, ports are still provided and used by overlay protocols for communication in DTN environments. Next, sequencing is done the same way as in TCP, with the exception that sequence numbers are assigned to *message bundles* rather than to individual packets. Connection establishment, on the other hand, is impossible in such environments due to the primary assumption of the absence of an end-to-end connection. The only remaining services, therefore, are congestion control and reliability. Congestion control is a more challenging function to deploy because propagating live congestion-related information across DTN environments is hard. This difficulty is due to the unstable nature of DTN environments. We mainly focus on reliability, a service critical to many of the applications that run in DTN environments.

In this chapter, we introduce four different end-to-end reliability approaches for DTNs. We use DTMNs as a representative DTN over which we study these reliability approaches. First, *hop-by-hop* reliability depends only on sending acknowledgments along every hop in the path. Second, *active receipt* achieves reliability by delivering an *active* end-to-end acknowledgment over the DTMN. Third, *passive receipt* reliability implicitly sends an end-to-end acknowledgment through the network. Fourth, *network-bridged receipt* sends an acknowledgment over another network that exists in parallel to the DTMN. With the multiple devices people currently carry, we can use other parallel networks, such as cell networks, as network bridges to transmit acknowledgements or other control-related information.

We evaluate these reliability approaches in DTMNs under various network conditions via simulations. Our goals in this study are to examine the impact of these reliability approaches, understand the tradeoffs between them, and open the way for further work in transport layer issues in delay tolerant networks.

5.2 Reliability Approaches

We present in this section the four reliability approaches that we study in this chapter. First, we discuss the most basic reliability approach for DTMNs, which is *hop-by-hop*. Afterwards, we talk about two different approaches for delivering an end-to-end acknowledgement over a DTMN. These approaches are *active receipt* and *passive receipt*. Finally, we propose a novel modification to the typical DTMN architecture by introducing the idea of *network-bridged receipt*.

5.2.1 Hop-by-hop

Hop-by-hop reliability was first introduced in classical DTNs [34]. The idea there, however, was to deliver a message across a given region on the path to the destination, where each region represents a hop. Gateways at the edges of these regions act as custodians and take the responsibility of reliably delivering message bundles across the region [35]. Therefore, there is no end-to-end acknowledgments in these cases; the source only knows whether the next gateway received the message or not, and assumes the gateway will take care of the rest. We build on this idea, and use it as the base reliability approach for DTMNs.

We apply hop-by-hop reliability, however, differently in DTMNs. With the extreme hostility and mobility assumed in DTMN applications, each node in the network acts as a region *and* a gateway with respect to the DTN architecture. Therefore, any exchange of messages between nodes is acknowledged, and all nodes are assumed to reliably forward the message.

The operation of hop-by-hop reliability in DTMNs is illustrated in Figure 5.1. The source, S, sends a message, M, to the ultimate destination, U, with the aid of forwarder nodes, F. Each time M is *successfully* delivered to any node, an acknowledgment, A, is then sent back to acknowledge the receipt of M. The forwarder nodes along with the source node try to infect as many nodes as possible according to their willingness level. Given enough time and mobility, S assumes that M will eventually reach U. Even though hop-by-hop does not ensure end-to-end reliability, it has the advantage of minimizing the amount of time M remains in S's buffer. This is because S does not need to wait for any end-to-end acknowledgment. We use hop-by-hop as the base approach over which we build the other end-to-end reliability approaches.

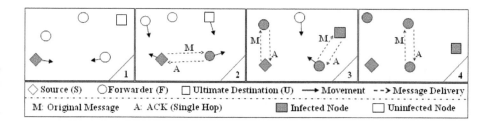

Figure 5.1: The operation of hop-by-hop reliability in DTMNs.

5.2.2 Active Receipt

While the hop-by-hop approach ensures some level of reliability, it does not ensure end-to-end reliability. This limitation could be a problem in cases where failures, such as the destruction of a node in a battlefield, or the breakdown of a node in a disaster rescue operation, are likely to occur. In such cases, some form of added end-to-end reliability is required. We overcome this drawback of the hop-by-hop approach by introducing the *active receipt*.

Active receipt is basically an end-to-end acknowledgment, which we call a *receipt*, created by U after it receives M from S. This receipt is *actively* sent back to S. By "actively", we mean that nodes treat this receipt as a new message that needs to be forwarded.

We demonstrate the operation of active receipt in Figure 5.2(a). The first snapshot starts at the time when U has just received M, with most of the F nodes already infected with M. U then creates the active receipt, R, which is forwarded

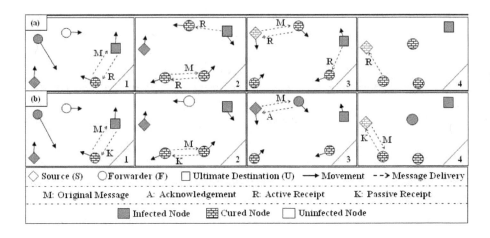

Figure 5.2: Demonstrating and comparing (a) active receipt and (b) passive receipt reliability approaches in DTMNs.

through the F nodes until it reaches S, shown in the third snapshot of Figure 5.2(a). Throughout this process, we observe how R *cures* the infected nodes in the network by stopping their transmission of M. R is also cached according to the nodes' willingness levels to prevent re-infection of M. Even though this cure eventually stops the epidemic spread of M through the network, R itself starts to spread epidemically until some timeout or TTL value. The cost of carrying and transmitting R, however, is less than M due to the small size of R.

5.2.3 Passive Receipt

While active receipt offers end-to-end reliability, its cost in many situations is high. This high cost is because active receipt reaches a point where two messages, rather than one, are infecting nodes in the network. Therefore, we introduce *passive receipt*, which ensures end-to-end reliability, without the incurred cost of active receipt. The idea is to have an implicit/passive receipt, instead of an active one, traverse the network back to S.

We use Figure 5.2(b) to help clarify the operation of passive receipt. The first snapshot, similar to Figure 5.2(a), starts at the time when U just received M. However, instead of generating a new active receipt, R, an implicit kill message, K, is sent to the infected node to stop it from sending M. The idea is that K is sent by the cured nodes (or U) *only* when they are encountered by one of the infected nodes trying to pass M on to them. In other words, cured nodes do not actively send K messages, they simply wait for active infected nodes to come in their way and stop them from sending M.

The operation of the passive receipt is better understood when compared to active receipt, as illustrated in Figure 5.2. The first difference is shown in both second snapshots, where in Figure 5.2(a), R is actively sent to an infected as well as an uninfected node. In the case of passive receipt shown in second snapshot of Figure 5.2(b), however, K is only sent to the infected node *after* this infected node had tried to pass M to a cured node.

This reduction in cost introduced by the passive receipt approach is not free when compared to active receipt. Even though an end-to-end receipt is received by S in both cases, S receives the end-to-end receipt more rapidly in the case of active receipt. When using the passive receipt, K is received by S at the fourth snapshot, as opposed to receiving R at the third snapshot using active receipt. The reason for this difference in receipt arrival time is that with the active approach, R spreads rapidly in the network, which helps it reach S more quickly than the passively spreading K. This passiveness also results in having infected nodes in the network take a longer time to be cured, as shown in the fourth snapshot in Figure 5.2(b). This means that the chances of having some infected nodes still trying to send M after S received a receipt, is higher in the passive receipt approach than the active receipt.

5.2.4 Network-Bridged Receipt

We now introduce a new assumption to the DTMN architecture that enables us to create another reliability approach. This assumption is based on the widespread use of cell phones. We propose exploiting the availability of the cell network by using it as an alternative path for our communication protocol. While such a network does not have the required bandwidth for delivering large bundled messages, it *could be* used for transmitting lightweight control information. Therefore, we use this cell network only for transmitting an end-to-end receipt from the destination back to the source.

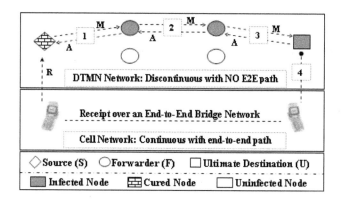

Figure 5.3: The network-bridged receipt reliability approach.

This idea is illustrated in Figure 5.3. We note that all nodes in the network are capable of mobility, however, for clarity, we do not include mobility in the figure. The cell network acts as a bridge between nodes in the DTMN. The cell network is characterized by its continuous end-to-end, low bandwidth connections. The DTMN network, on the other hand, is characterized by its discontinuous non-end-to-end, high bandwidth. In such a setup, large messages, M, are typically transmitted from S over the DTMN using the base hop-by-hop reliability approach until it reaches U. The end-to-end network-bridged receipt, R, would then be transmitted over the cell network instead of the DTMN. If we assume that other nodes in the network also have access to the cell network, R could then be transmitted to these nodes. The result is a very rapid cure for all infected nodes in the network.

The advantage of the network-bridged approach is to reduce the round trip time between nodes S and U roughly by half. Consequently, the message is dropped faster from the queue in A since the receipt arrives faster. The drawback, however, lies in the assumption itself: the added complexity of bridging the DTMN network with the cell network. We believe, however, that the inter-connection of these two networks is a likely possibility in the future.

5.3 Evaluation

The primary goal of our evaluation is to compare the performance and examine the tradeoffs between the reliability approaches described in Section 4. We first describe our simulation setup and environment. We then summarize the outcomes of an extended set of simulations we conducted. We only present a subset of our results that most clearly allows us to show the tradeoffs between our reliability approaches.

5.3.1 Simulation Environment

We conducted our simulations using the GloMoSim network simulator [104]. We added an overlay layer that handles message bundle relaying and implements the reliability approaches that we have described. We use a *modified* random way-point mobility model that avoids the major problem of node slow down in the conventional random way-point model. We believe this model closely approximates the scenarios with which we are concerned, such as battlefields or disaster rescue operations, due to their hostility and unpredictable movement. The node speed ranges between 20 to 35 meters per second, and the rest period is between 0 and 10 seconds. We examined other ranges as well, and they produced similar results with respect to our reliability approaches. Every point in our results is taken as an average of ten different seeds.

The major parameters used in our simulations are summarized in Table 5.1. The *Terrain* is the area over which the *Number of Nodes* are scattered. *Simulated Time* represents the amount of time the simulations run. The *Beacon Interval* is the period after which beacons are sent. A "beacon" is simply a signal emitted by all nodes to search for other nodes in the network as well as to announce its location. The *Times-To-Send* (TTS) is the number of times a node will successfully forward a message to other nodes in the network. *Retransmission Wait Time* represents the amount of time a node remains idle after successfully forwarding a message to another node. When the retransmission wait time expires, the node then tries to resend the same message. We mainly use TTS to represent the *willingness* of the nodes to participate in message relaying. Finally, the *reliability approach* parameter represents our four different acknowledgement schemes.

Table 5.1: Reliability approaches simulation parameters

Parameter	Value Range	Nominal Value
Terrain	$10km^2$ to $50km^2$	$10km^2$
Number of Nodes	10 to 250	100
Simulated Time	1hour to 24 hours	6 hours
Beacon Interval	0.5sec to 50sec	1sec
Times-To-Send	1 to 50	10
Retransmission Wait Time	0sec to 500sec	50sec
Reliability Approach	Hop-by-hop, Active, Passive or Network-Bridged	N/A

We consider three main metrics in evaluating our reliability approaches. The first metric is *Cost*, which is the total number of messages sent by all nodes in the network. The second metric is *Queuing Time*, which is the average time a message remains in the sender node's queue before it is dropped. The third metric we consider is *Delivery Ratio*, which is the percentage of messages delivered. We choose to focus on the first two metrics since delivery ratios in DTMNs simply depend on the time ceiling set for message delivery, i.e. given enough time, all messages will eventually be delivered.

5.3.2   Results

We present a summary of the extended set of simulations, along with a subset of our simulation results, which clarify and support these findings. All the results are shown for a single sender node sending one message to a single ultimate

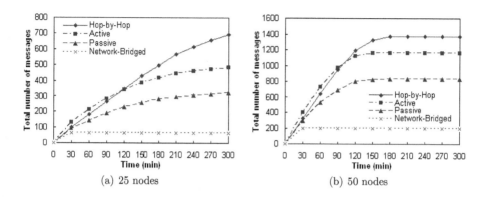

Figure 5.4: The cost of the reliability approaches over time in DTMNs with different node densities. Graphs (a) and (b) represent 25 and 50 nodes, both with a TTS of 10.

destination. The purpose of our simulations is twofold. First, we hope to better understand how different reliability approaches behave when run in a DTMN. Second, we want to understand the tradeoffs between these approaches.

Generally speaking, the network-bridged receipt incurs the least cost when compared to the other approaches. The highest cost, on the other hand, occurs with the hop-by-hop approach. The cost of the active and passive receipts fall in between, with active receipt being relatively more expensive. These observations are supported by Figure 5.4 and Figure 5.5, which demonstrates the cost of each reliability approach in terms of the total number of messages sent. We measure this cost under different network densities, 25 nodes in Figure 5.4(a), 50 nodes in Figure 5.4(b) and 100 nodes in Figure 5.5(d), as well as different willingness levels, times-to-send is set to 5 in Figure 5.5(a) and 10 in Figure 5.5(b). One

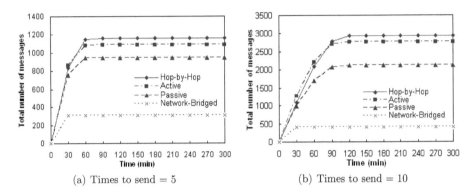

(a) Times to send = 5   (b) Times to send = 10

Figure 5.5: The cost of the reliability approaches over time in DTMNs with different willingness levels. Graphs (a) and (b) represent TTS of 5 and 10, both with 100 nodes.

interesting observation is where the cost of the active receipt is the highest until it is eventually exceeded by the hop-by-hop approach. This result is because after the message reaches the ultimate destination, we now have two messages infecting the network, which creates this large cost. Eventually, however, the receipt cures those nodes infected with the original message and is itself cured after reaching the source node. We note also that changes in node density or willingness levels have minor impact on the *relative* performance of our reliability approaches.

Even though the performance of the reliability approaches is relatively similar over different network densities, other aspects, such as the rate of message spreading and convergence, vary. This result is particularly evident in the difference in the Y-axis scales of Figure 5.4 and Figure 5.5. Generally speaking, the messages spread faster in denser networks. This observation can be seen by the sharper

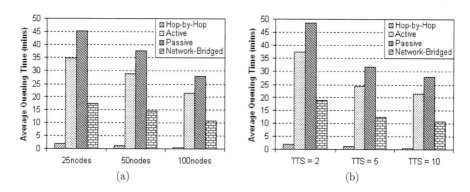

Figure 5.6: he impact of (a) the number of nodes, and (b) the times-to-send on the average queuing time of a message at the sender node.

increase in the total number of messages in the case of Figure 5.5(b) when compared to Figures 5.4(a) and Figure 5.4(b). We compare Figure 5.5(b) with Figure 5.4 since the former measures the cost over a 100 node network with the same TTS value of 10 as that used in Figure 5.4. Alternatively, the network heals faster in denser networks. This result is shown in the faster convergence of the lines in Figure 5.4(b) when compared to those in Figure 5.4(a). This convergence leads to a steady horizontal line because the network reaches a point of saturation where it no longer needs to forward the message.

Regarding the average queuing time, the results show that the hop-by-hop approach has the lowest value. This low value is because the source node does not wait for any end-to-end acknowledgement to be received, and therefore, drops the message from its buffer after forwarding to other nodes in the network. If end-to-end reliability is required, the best approach in terms of minimal queuing time is the network-bridged approach. Figure 5.6 supports these observations by illustrating the average queuing time of a given message with respect to our reliability approaches under (a) different densities, and (b) different willingness levels. The other interesting fact Figure 5.6 highlights, is that active receipt has less queuing time than passive receipt. This fact offers a tradeoff for the extra cost incurred in the active receipt when compared to the passive receipt approach. The reason for this result is due to the active way in which the receipt is sent when compared to the passive approach. The active approach results in the receipt reaching the source faster, but at a higher cost.

Figure 5.6 also shows that the tradeoffs between the reliability approaches is generally similar over different densities and different willingness levels. The primary difference is that the overall queuing time of all the reliability approaches decreases as the network density or willingness levels increase. This result is because in denser networks, or when nodes are trying harder to forward a message, the overall end-to-end delay decreases. This decrease in delay consequently leads to smaller queuing time.

5.4 Summary

In this chapter, we have considered transport layer issues, specifically reliability, over a special class of DTNs knows as DTMNs. We introduced four different reliability approaches: hop-by-hop, active receipt, passive receipt, and network-bridged receipt. We have investigated and evaluated these approaches via simulation. Overall, we discovered that the choice of the most suitable reliability approach depends on the expected complexity of the underlying DTMN. For example, the hop-by-hop is the simplest, while network-bridged is the most complex. Also, the priority of cost versus delay governs the choice between the active and passive receipt.

We consider the work in this chapter a next step in thoroughly investigating transport layer issues in DTNs in general. Our future work in this area, therefore, is to apply these approaches to DTNs in general, and see how they might be modified and applied to other DTN architectures. Also, we intend to address other transport layer issues, particularly, congestion control.

Chapter 6

Intermittent Mobile Connectivity

6.1 Introduction and Motivation

So far in the dissertation, we have been focusing our attention on various
delay and disruption tolerant applications that experience relatively large delays
and disruptions. Nodes in these networks occasionally experience connectivity
intervals, but mostly remain disconnected. However, there is a parallel rising
demand for another class of mobile applications that experience repetitive short
disconnections, such that the total connection time is larger than the disconnection
time.

The response to this demand has been the development of mobile applications
and devices that can be used by in-motion users. However, as users move between
connection points, they experience bursts of network connectivity interspersed
with either weak or non-existent signals. A recent study finds that mobile devices

can move at speeds of 75 mph and still experience periods of connectivity with high throughput and low loss [37]. However, most, if not all, current applications are not designed to take advantage of these short network connectivity bursts. An insufficient amount of data is exchanged before a disconnection occurs, and often must be re-gathered the next time a connection is present. The intermittent connectivity in such scenarios leads to large latencies, user frustration, and possibly even complete application failure.

This user frustration is more fully appreciated with the following scenario. A mobile user takes a bus to work every morning. At a certain point in his commute, this user notices that his laptop has detected a signal from a nearby wireless Access Point (AP). The user opens a web browser and connects to www.cnn.com to read the morning news. He reads the blurb for the main story and clicks the link to read the full text. However, by this time the bus has moved past the AP's signal range and the user receives a Page Not Found error. Although the user's laptop was likely connected long enough to receive a significant amount of data, part of this time expired before the user realized a connection existed, and more time was wasted as he read the main story's blurb. A system that quickly reacts to the acquisition of a signal, and utilizes the full connection period, would greatly enhance the user's experience in this case. If the system knows that the user enjoys reading the morning news on her way to work, it can proactively gather this data in the background whenever a connection is available. The full text of the main news story will then be awaiting the user when she wants to read it.

While some researchers have developed solutions that hide the ill-effects of intermittent connectivity. The majority of these proposed solutions focus on reacting intelligently to disconnections after a request has been made [7], [62], [75]. Possible reactions include either caching requests [26], [57], [72] or maintaining high-level connections [71], [76]. In all of these solutions, requests made during times of disconnection wait to be serviced until connectivity returns. In addition, the mobile devices in these solutions are required to open separate connections to each application server the user wishes to contact. The system we present in this chapter avoids both of these drawbacks.

We present and develop DBS-IC, a Data Bundling System for Intermittent Connections, which takes advantage of short connection periods to enhance the experience of mobile users experiencing intermittent connectivity. A Stationary Agent (SA), located on a stationary device with a stable connection, collects data the user has specified will be needed in the future. This data can be heterogeneous: data from web servers, email servers, and other file servers. The SA then groups this data together into a single package, or *bundle*. Afterwards, the SA opportunistically sends this bundle to a Mobile Agent (MA), residing on a mobile device, whenever a connection is present. Once the bundle is successfully transferred to the MA, the user can view the data at any time, including times of disconnection. In this way, our system hides the underlying instability of the connection.

DBS-IC efficiently utilizes available bandwidth using a combination of multiple techniques. First, our system forms a single connection between the MA and SA to send heterogeneous data, thereby eliminating application-specific connection and request times. In the scenario above, the user may want to check her email after reading the morning news. DBS-IC, therefore, sends the user's emails in the same bundle as the web data from CNN, relieving the user of the need to contact these servers separately. Second, after the first copy of a data bundle has been sent to the MA, DBS-IC bundles and sends only data *updates* in an effort to eliminate unnecessary re-transmissions. Referring to our scenario, after the email and web data has been sent to the user, only new emails and updated web content will be sent in the future.

With respect to bundling, we discuss various bundling schemes and address the transfer latency and data staleness that can arise from overly large bundles. For example, if the user's laptop is experiencing extremely short connections, the data may be out-of-date when the transfer finally completes. To counteract these problems, we introduce *mini-bundles* to expedite the transfer of data that is immediately viewable by the user and to keep data current. Instead of sending the user's email and the large amount of CNN web data all at once, we can send a chunk at a time so that the data is incrementally available. We examine different approaches for creating these mini-bundles, including data type, size, and priority.

We fully implement DBS-IC in order to evaluate its performance. We choose to implement DBS-IC, rather than simulate the system, in an attempt to obtain more realistic results under unpredictable wireless network conditions. We evaluate the performance of our implemented system in different intermittent connectivity scenarios, and compare the results to existing data retrieval methods. Results of live tests are excellent, showing that DBS-IC efficiently utilizes bandwidth to opportunistically deliver data to the user before disconnections occur. We find that mini-bundles further enhance our system, delivering viewable data to the mobile user significantly faster than traditional retrieval protocols such as the Hyper Text Transfer Protocol (HTTP).

6.2   DBS-IC

In this section, we present an overview of our Data Bundling System for Intermittent Connections (DBS-IC) and its associated protocol, the Mobile Proactive Transport Protocol (MPTP). We first explain the major components and overall operation of our system. We continue with a discussion of several critical issues such as authentication, bundling, and handling data updates. Next, we introduce mini-bundles, and show different methods by which they can be constructed. Finally, we conclude this section with a discussion of additional design considerations.

6.2.1 System Components and Operation

There are two major components that comprise our Data Bundling System for Intermittent Connections (DBS-IC). These components are a Stationary Agent (SA) and a Mobile Agent (MA). The SA is located on a stationary device that has a stable connection to the Internet, such as a user's desktop computer. The SA gathers various forms of data from different sources, such as web, email, and file servers. The MA is located on an in-motion mobile device which moves between wireless Access Points (APs) or mesh routers and therefore experiences intermittent connectivity. A Mobile Proactive Transfer Protocol (MPTP) connection is created between the SA and MA whenever the MA enters connectivity range. We created MPTP to help provide authentication, lower connection overhead, and provide more bandwidth for data transfers. Details of this protocol are not included due to space limitation.

The DBS-IC architecture and a basic operation scenario are presented in Figure 6.1. Operation begins with an initial configuration step in which the user interacts

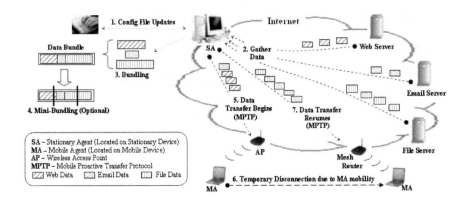

Figure 6.1: The Data Bundling System for Intermittent Connectivity (DBS-IC) architecture and operation scenario.

with the SA, specifying web, email, and file servers from which the agent should

gather data (Step 1). The SA then contacts the specified application servers

and gathers the user-requested data (Step 2). This step is periodically repeated,

based on a user-configurable update frequency, to keep the data current. After

gathering all the requested data, the SA then bundles these pages, emails, and

files into a single group. The SA is then ready to send this bundle to the MA as

soon as the SA is contacted (Step 3). An alternative to bundling all the data into

one large bundle, called *mini-bundling*, can optionally be performed at this point.

With mini-bundling, the gathered data is structured into multiple smaller bundles,

which can each be autonomously sent to the MA (Step 4). The methodology and

benefits of mini-bundling are discussed later.

At this point, the SA is ready to transmit data to the MA. When the MA comes within range of an access point, the MA contacts the SA. In response to this signal, the SA begins to send the data it has previously gathered and bundled (Step 5). However, in this example, the MA moves out of range and loses its connection in the middle of the transmission (Step 6). We employ heartbeat messages, sent from the MA to the SA at regular intervals, to help the SA recognize disconnections. When the MA loses connectivity, its heartbeat messages no longer reach the MA. As a result, the SA learns that a disconnection has occurred. The SA stops the transmission and waits until it is again contacted. When the MA moves back into connectivity range, it again contacts the SA. The SA then resumes sending the data from the last byte the MA received (Step 7). As soon as the SA finishes sending a bundle, the MA has the opportunity to send configuration file updates. This cycle repeats indefinitely: as long as the SA is running, updates will be gathered and bundled, waiting to be sent to the MA.

6.2.2   Authentication, Bundling, and Updating

We now discuss in more detail the features of DBS-IC that allow the system to efficiently utilize short connection periods to transmit data. Recent study findings show that even up to speeds of 75 mph, an intermittently connected mobile device experiences connection periods with high throughput and low loss

141

[37]. The two main factors prohibiting meaningful data transfer at these speeds are (1) lengthy connection and authentication times, and (2) multiple application-required request cycles (such as multiple HTTP GET requests). Our system solves both of these problems by providing a low-overhead authentication scheme, and by bundling data to avoid multiple connections from the mobile device.

In the basic DBS-IC operation scenario, presented above, there are some important features that solve the two problems discussed in the preceding study. First, authentication via a username and password is not needed the second time the MA contacts the SA. Instead, the MA will use a unique session identifier that the SA assigns to it. This simpler authentication technique lowers overhead and provides the system a larger percentage of connection time devoted for actual data transfer. Another important feature of our system is how it delivers data to the MA. Once the MA receives a complete copy of an initial bundle, the SA will send only updates to this data in the future. This technique will save bandwidth and unnecessary re-transmission of data the MA has already received.

While the SA continues to gather data updates even when the MA is disconnected, the SA does not apply these updates to any of the data that has been partially sent to the MA. The SA instead sends the MA a complete copy of the data before sending updates, even if this means sending stale data. Two other possible methods for handling updates are possible. First, if the MA is not currently connected, and an update is gathered, the SA could start sending from the beginning of this updated data when the MA re-establishes a connection. Second, the SA could update only those pages, emails, or files that have not yet been sent to the MA. The problem with the first alternative is that there is a risk the user will never receive a complete copy of the data, instead receiving only the beginning of multiple versions of data. This partial data cannot be unbundled and viewed by the user until all the data has been received. The second alternative is difficult because the SA does not know exactly how much information the MA received before being disconnected. In the current implementation, the SA only ascertains this information once the MA reconnects and sends a byte count to the SA. Hence, the SA cannot know exactly what data can safely be updated.

In addition to the stale data problem discussed above, large bundles can result in overly long data availability latencies. More specifically, our system can enter a situation in which there is a significant amount of data on the MA that cannot be viewed by the user. Consider the following scenario. The MA connects to the SA and receives half, or even 90%, of a large bundle of data before losing connectivity. The MA cannot re-connect to the SA for another hour. Because it has not received the entire bundle, the user cannot view any of the data already received. The entire bundle is needed before it can be unbundled, and, in the case of an update, patched with the data the MA already has. To solve this problem, we introduce mini-bundles.

6.2.3 Mini-Bundles

Mini-bundles are pieces of the complete data bundle. By dividing the data bundle and transmitting smaller mini-bundles, there is a higher probability that the MA will receive the entire mini-bundle before experiencing a disconnection. And receiving the entire mini-bundle, as discussed earlier, is a prerequisite for the MA to view any data. However, due to the communication overhead that takes place after data is transferred, there is a chance mini-bundles could hurt system performance. A balance between the number of mini-bundles and the size of each bundle is needed. Mini-bundles can be created based on various criteria, such as the following:

1) *Application Type:* Data of one type is bundled separately from data of another type. For example, one mini-bundle is composed entirely of email data, another contains only web data, and a third consists of file data.

2) *Priority:* Mini-bundles are created based on user-specified priority values. One mini-bundle consists of the data of the highest priority, another of lesser priority, etc.

3) *Size:* Regardless of the content of the bundle, the bundled data is divided into equally sized mini-bundles. This means that each mini-bundle could contain similar or different types of data.

When used alone, each of these techniques has benefits and drawbacks. We employ a merged technique in which mini-bundles are formed primarily based on size, secondarily on priority, and finally according to data type. This merged technique delivers the data the user wants most, while regulating the mini-bundles so they are approximately the same size. The size equality of mini-bundles ensures that one greedy mini-bundle does not monopolize the connection time, thereby defeating the purpose of mini-bundles. Regardless of the method that is used to create them, each mini-bundle is self-contained. More specifically, a piece of data that the user wants (for example, a web page or a set of emails from a single email server) will never be divided between mini-bundles. Therefore, after the transfer of a mini-bundle is accomplished, the user has a complete, viewable subset of the bundled data.

6.2.4 Other Design Considerations

In our current implementation of DBS-IC, all MPTP connections are built on top of TCP. Because TCP provides reliability and in-order delivery, MPTP itself is not designed to provide these features. We justify using TCP with two reasons.

First, reliability and in-order delivery are imperative for our system; every packet of a transmission must be received. We briefly attempted to develop a system using UDP connections, but without additional services built into MPTP, UDP fails. As soon as one packet is lost, the system breaks. To fix this fragility, we would need to add more frequent data acknowledgments and timeouts, both of which would slow the system.

Second, we justify our decision to use TCP based on the results of Gass et al.'s previous study, which finds that TCP bulk data transfer to an in-motion mobile device actually achieves much higher throughput than UDP [37]. Based on our experience with UDP, and the estimated overhead of making UDP reliable, all of the MPTP connections between the MA and SA are built on top of TCP. Possible future work in this area would be to examine the benefits of modifying TCP window size on the throughput of our system. However, we show in the evaluation section that our system still achieves high throughput without modifying TCP, even when only short connection periods can be guaranteed.

Another design consideration of DBS-IC is the initial configuration step in which the user specifies the data they want delivered to their mobile device. Instead of requiring the user to complete this step, the SA could incorporate a smart gathering agent which retrieves data based on recent user browsing trends. We chose not to focus on the smart gathering agent at this stage. Instead, we focus on bundling and sending data proactively to examine how user experience is affected. Adding a smart agent mainly reduces the burden on the user by automating the decision for which data needs to be gathered. This improvement is left for future work.

6.3 Evaluation Setup and Environment

In the current and following sections we present the evaluation of DBS-IC. We use our evaluation to accomplish two main goals. The first goal is to compare the performance of DBS-IC in situations of intermittent connectivity to the performance of traditional retrieval methods. Our second goal is to examine the impact of various parameters on the performance of DBS-IC. We discuss the evaluation setup and environment, followed by a set of results that fulfill our designated goals.

We fully implement DBS-IC and test our implementation in various ways. We chose to implement our system, as opposed to simulating it, to obtain more realistic results. We perform our tests in the lab, using the results gathered in previous driving test studies [37], [75] to accurately and realistically choose some of our parameters as shown in Table 6.1. The SA is located on a lab machine that has a stable connection to the Internet, with an unloaded 100 Mbps full-duplex

Table 6.1: DBS-IC Evaluation Parameters

Parameter	Value Range	Nominal Value
Bundle Size	1 MB to 30 MB	20 MB
Connection Duration	15 sec to 100 sec	20 sec
Disconnection Duration	10 sec to 30 sec	10 sec
Number of Mini-Bundles	1 mini-bundle to 5 mini-bundles	3 mini-bundles
Mini-Bundling Technique	Size, Priority, Data Type, Merged	Merged
Prefetched Data	0% to 100%	100%
Intermittent Connectivity Model	Walking, Downtown, Suburb, Highway	Suburb
Round Trip Time	10 ms	10 ms

switched Ethernet connection. The MA is located on an intermittently connected laptop with an 802.11b network card.

The SA gathers and bundles three types of data in our tests: web pages, emails, and files. There is an average round trip time of 10 ms between the MA and the SA. We expect round trip time will impact our system when the distance between the SA and MA is greater than the distance between the MA and the individual application servers. If it is significantly faster for the MA to contact the application servers than for the MA to contact the SA, our system may not be as useful; however, the bundling and single connection features of DBS-IC will still offer a significant performance improvement.

Throughout our evaluation, we examine different *intermittent connectivity models* to see how our system performs under varying mobile situations. Each model is characterized by a unique combination of connection and disconnection durations, as shown in Table 6.1. Gass et al. [37] show that a mobile device traveling 5 mph experiences approximately 100 seconds of efficient connection time while passing a single access point, discounting the lossy entry and exit phases discussed by Ott and Kutscher [75]. Similarly, a mobile device experiences about 40 seconds of connectivity at 35 mph and 15 seconds at 75 mph. Using these numbers as a guide, we recognize the following four intermittent connectivity models:

1. *Downtown Walking Model:* This model is characterized by connections of 120 seconds and disconnections of 20 seconds. It simulates the experience of a mobile user walking past access points or mesh routers in an urban area. We do not focus heavily on this model since the lengthy connection periods are generally adequate to gather data using traditional retrieval methods like HTTP and FTP.

2. *Downtown Driving Model:* This model is characterized by connections of 40 seconds and disconnections of 15 seconds. It simulates the experience of a mobile user driving in slow traffic.

3. *Suburb Driving Model:* This model is characterized by connections of 20 seconds and disconnections of 10 seconds. It simulates the experience of a mobile user driving on surface streets without traffic.

149

4. *Highway Driving Model:* This model is characterized by connections of 15 seconds and disconnections of 30 seconds. It simulates the experience of a mobile user driving at 70 mph along a highway that passes periodic access points.

In addition to the intermittent connectivity model, we also vary characteristics about the data our system is transmitting throughout our tests. We define *bundle size* as the size of the compressed data that is physically transferred between the SA and the MA. This bundle is divided into a specific number of *mini-bundles*, which we vary from one to five to evaluate the improvement mini-bundles provide in delivery time versus the tradeoff of additional overhead. These mini-bundles are created based on the *mini-bundling technique*, which can be size, priority, data type, or the merged technique discussed earlier. The time it takes to make data available at the MA also depends on the percentage of *prefetched data*, the amount of data that was gathered by the SA before the MA connects to the SA. In the optimal case, the SA will have gathered 100% of the data before the MA connects, but we find that our system still performs well when this is not the case.

The metrics we use in our evaluation are the following:

User-Perceived Data Delivery Time: The time between when the user decides to view a piece of data and when that piece of data is viewable on the MA.

Data Throughput: The amount of data our system transfers between the SA and MA during connection periods. Our system is built to maximize this throughput by lowering connection overhead.

Data Staleness: The time difference between the latest version of a piece of data on the MA and the latest version on the SA. Our system hopes to deliver data updates in a way that will minimize data staleness.

6.4   Implementation Results

We divide our results section into five main parts. Initially, we analyze how quickly data becomes available to the mobile user when using our system as compared to using existing retrieval protocols. We then take a closer look at the throughput our system achieves and follow this by testing the performance enhancements that mini-bundles provide to our system. We then evaluate how DBS-IC performs using different intermittent connectivity models. Finally, we discuss the performance of our system when some or all of the data the mobile user wants has not been prefetched by the SA.

6.4.1   Data Gathering and Data Transfer

As stated earlier, the goal of DBS-IC is to opportunistically present the mobile user with data, so disconnections will have a less adverse effect on viewing data. Currently, protocols such as HTTP, FTP, and SMTP are used to gather web, file, and email data, respectively. These protocols were not designed with intermittent connectivity in mind; they have long connect, request, and timeout cycles. And in the case of HTTP, a new TCP connection must be created for each page requested. By gathering this data on a machine with a stable connection to the Internet, our system reduces the data transfer on the mobile device to an opportunistic bulk transfer of bundled data. Only one TCP connection is created when the mobile device gains connectivity, avoiding the connection overhead of each individual piece of data.

Our first test compares the time to gather data on the MA using traditional retrieval methods to the time to both gather the data on the SA and transfer it to the MA using our system. In both cases, the MA experiences intermittent connectivity based on the suburb driving model. We vary the total data size in this test, which means the compressed bundled data is smaller and varies depending on the type of data. In an attempt to keep the compressed data consistent, each bundle consists of 2/3 web data, 1/6 email data, and 1/6 file data.

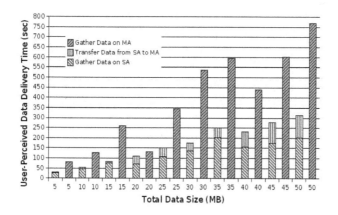

Figure 6.2: Data gathering vs. data transfer in DBS-IC.

Figure 6.2 shows that in all cases, the user-perceived data delivery time is reduced when our system is used. On average, the time is reduced by a factor of two. The large gathering times experienced when the MA uses traditional retrieval methods is caused by two main factors. First, the MA is experiencing disconnections every 20 seconds, which slows web data retrieval significantly. When retrieving web pages on the MA, we set a 10 second timeout value. With this timeout value, a retrieval or connection is automatically re-tried if a response is not heard within 10 seconds. This timeout simulates a user hitting the refresh button after the page has been trying to load for 10 seconds. Since the MA is experiencing 10 second disconnections, this 10 second timeout is a lower bound on usefulness. A lower timeout value will have no beneficial effect on performance since the MA will still be disconnected when a connection is re-tried.

The second reason for the longer retrieval time on the MA is due to the fact that the MA is located on a wireless device with an 802.11b network card. It therefore has inherently less bandwidth and throughput than a wired device. In this test, we add the data retrieval time on the SA to the transfer time from the SA to the MA. This combination means that in the worst case, when the SA has prefetched 0% of the data, our system still performs relatively well. In many cases, however, the SA will have proactively gathered this data, before the MA ever connects to the SA. In this situation, the user-perceived data delivery time consists only of the transfer time between the SA and the MA, reducing data availability times, on average, by a factor of 12 as compared to traditional methods.

6.4.2   The Impact of Mini-Bundles

The previous test helps show that our system delivers data the user wants faster than if the mobile user gathers it themselves using HTTP or similar existing retrieval protocols. However, if the user is gathering data, such as web pages, each page will be viewable as soon as it loads. While the user will not have the complete data for a while, they will have part of the data to keep them occupied. We now empirically examine the speedup in user-perceived data delivery time that our system obtains by utilizing mini-bundles.

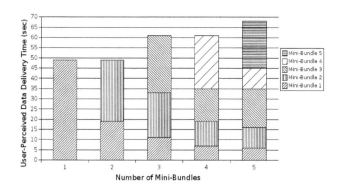

Figure 6.3: Impact of mini-bundling on data availability in DBS-IC.

We first test mini-bundles by holding connection duration and bundle size constant, varying the mini-bundle count from 1 to 5. We again follow the suburb driving model. In this test, all mini-bundles are of equal size. More specifically, the SA gathers and bundles 30MB of data, then sends this 30 MB of data to the MA in a varying number of mini-bundles. The SA first sends the data in one 30 MB mini-bundle, then two 15 MB mini-bundles, then three 10 MB mini-bundles, and so on.

Figure 6.3 shows that as the number of mini-bundles increases, the user-perceived data delivery time of *some* data is decreased. Compared to sending the data in one mini-bundle, the user can view data in half the time with two mini-bundles. Of course, only half of the total data is viewable at this time. A less desirable effect of mini-bundles is the added overhead. There is extra com-

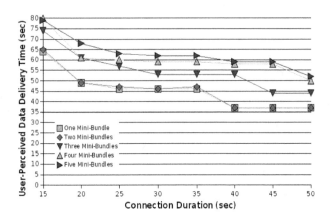

Figure 6.4: Impact of connection duration on data availability in DBS-IC.

munication costs with each additional mini-bundle, since the SA must prepare the MA for each mini-bundle it sends, and the MA must acknowledge every mini-bundle it receives. The MA also has the opportunity to send configuration file updates between each mini-bundle, in an attempt to keep the data as up to date as possible. However, although the total data delivery time does increase, mini-bundles present the mobile user with viewable data faster than sending the data in one bundle. And since mini-bundles can be arranged by priority, they are an efficient way to opportunistically deliver to the user their most important data more quickly.

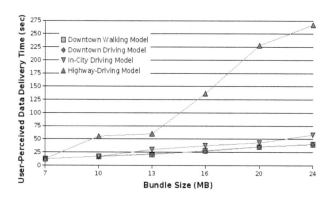

Figure 6.5: Impact of connectivity model on data availability in DBS-IC.

6.4.3 Intermittent Connectivity Model

We next evaluate the performance of mini-bundles when our system experiences different connection durations. To compare mini-bundle overhead, we only look at total data transfer times in this test; we do not take into account the partial data delivery improvements provided by mini-bundles. We also hold disconnection time constant at 10 seconds in order to isolate the effect that different connection durations have on our system. In the previous test, we saw that using additional mini-bundles resulted in a longer overall data transfer time. We observed that this time increase was due to the fact that the extra processing costs associated with more mini-bundles consumed a portion of the short connection periods available.

In Figure 6.4, we see that the same pattern holds, regardless of the connection duration. Each additional mini-bundle adds some processing overhead, increasing the user-perceived data delivery time for the whole data in all situations. Therefore, mini-bundles should only be used when some part of the data is more crucial to the mobile user than other parts. If the user needs to receive the entire data together, using only one bundle remains an efficient choice.

While holding disconnection periods constant in the above test is a nice way of isolating the effect of connection duration, it is not realistic. Figure 6.5 plots the user-perceived data delivery time against bundle size for the downtown walking, downtown driving, suburb driving, and highway driving models. The trends for all models are very similar, with the highway driving model having the largest user-perceived data delivery time due to its short connection and long disconnection periods. The long disconnection periods in this model also lead to the highest levels of data staleness. However, with such short connections, traditional retrieval methods will suffer worse than our system suffers, making DBS-IC a viable solution in all intermittent connectivity models.

6.4.4   Data Throughput

We have so far only vaguely examined data throughput between the SA and the MA. The transfer bandwidth that our system achieves warrants further examination. Therefore, we next examine the instantaneous throughput our system obtains while experiencing 15, 30, and 45 second connection periods. In this test, we hold bundle size constant at 30 MB and disconnection duration constant at 10 seconds. With no disconnections, our system transfers this 30 MB of compressed data in 59 seconds. We therefore do not examine the throughput obtained in situations where the MA experiences connection periods of greater than 45 seconds, since no disconnection would occur during data transfer. In these cases, our system remains beneficial if the 30 MB of compressed data can be transferred during a period of connectivity while the larger, uncompressed data cannot.

In Figure 6.6, we see that our system achieves instantaneous throughput of up to 5.24 Mbps. As expected, the throughput drops to 0 when a disconnection occurs. We can further see that the longer the connection period, the less time it takes to transfer the 30 MB of data to the MA. Our system delivers the data to the MA in 82 seconds when the MA experiences 15 second connections (Figure 6.6(a)); in 60 seconds with 30 second connections (Figure 6.6(b)); and in 58 seconds with 45 second connections (Figure 6.6(c)). Even in the worst case of 15 second connection periods, simulating a mobile device moving at 70 mph, our system averages a throughput of 2.92 Mbps. This throughput is the average over the entire data transfer, which includes two periods of disconnection. When discounting the disconnection periods, the throughput increases to an average of 4.14

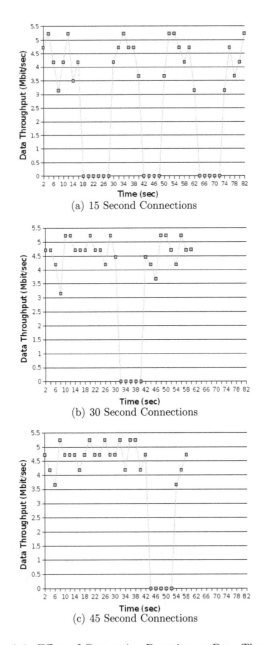

Figure 6.6: Effect of Connection Duration on Data Throughput

160

Mbps. As a comparison, the MA needed 535 seconds to gather 30 MB of web, file, and email data in Figure 6.2, resulting in an average throughput of 0.45 Mbps. By reducing the data transmission to a transfer of bundled data, our system achieves significantly more throughput than traditional retrieval protocols. This increased opportunistic data throughput means the mobile user will have viewable data faster, and that this data will remain viewable even during disconnections.

6.4.5   Prefetched Data

In the above set of tests, with the exception of the first, we assumed that the data the user wishes to view has been completely prefetched at the SA. Therefore, we have been considering the transfer of the bundled data as the only factor keeping the mobile user from viewing it. However, there will be added overhead if the data has not been completely prefetched before the user wishes to view it. Figure 6.2 showed that even when 0% of the data has been prefetched, our system transfers all the data to the mobile user faster than the user could gather it using traditional techniques. Our system will experience this case of 0% prefetched data when the user modifies the configuration file on the MA. After this change is made, the MA will send the updated configuration file to the SA, the SA will immediately gather and bundle any newly requested data, and send the bundled data back to the MA. Since our system performs well in this extreme case, we can easily modify Figure 6.2 for different percentages of prefetched data, and to see that our system performs well in all cases.

6.5 Summary

In this chapter we have presented a Data Bundling System for Intermittent Connections (DBS-IC), a system which deals with intermittent connectivity by proactively delivering data to the mobile user. DBS-IC is comprised of a Stationary Agent (SA), located on a machine with a stable connection to the Internet, and a Mobile Agent (MA), located on an intermittently connected mobile device. By confining the gathering of web, email, and file data to the SA, DBS-IC reduces the data transfer on the mobile device to a bulk TCP data transfer, which allows our system to utilize available bandwidth extremely well. We find that our system can make data available to the mobile user up to 20 times faster than if the data were gathered on the mobile device itself, even when the mobile device is only experiencing connection periods of 20 seconds at a time.

We believe that DBS-IC is a solid step to improve mobile users' experience in the face of intermittent connectivity. However, there are several areas for future work. A valuable future improvement to our system would be the addition of an intelligent gathering and bundling agent. This agent would be located on the SA and would use past viewing trends to dynamically decide which data the user might need in the future. Our system could further be extended to handle interactive data, caching user requests during times of disconnection. This extension would be especially useful with interactive web pages that require user input, and with newly composed emails that the user wishes to send. Built on top of our work, these improvements would help make the in-motion mobile user's experience almost equal to that of a stationary user's.

Chapter 7

A Parallel Network Architecture for Challenged Networks

7.1 Introduction and Motivation

Despite the fact that the challenges posed by delay and disruption tolerant networks have been addressed in different ways for various applications, we observe a general trend in the solutions and architectures presented so far. This trend lies in the fact that all of these solutions are based on the idea of operating over a single homogenous network, or sequential heterogeneous overlay networks.

In parallel to this trend, we have begun to witness the inevitable convergence of different networking technologies. This convergence occurs by providing communication alternatives to users through carrying multiple devices, or a single device, with access to multiple networks [25], [10]. However, network protocols that operate on these devices are primarily designed to operate over a single network at a given time, or multiple channels within the same network. Most, if not all, current approaches for providing inter-operability between heterogeneous networks rely on a high level overlay protocol that performs protocol translation. These overlays, however, are usually at network gateways rather than endpoints. We believe we can exploit the current and expected future convergence of networking technologies to better serve challenged networks.

Based on our vision, as well as the current trends in challenged networks research and networking technologies, we propose the *Parallel Networks (ParaNets)* architecture. The idea behind parallel networks is to provide an architecture over which network protocols, developed for challenged networks, can seamlessly utilize multiple heterogeneous networks in parallel. Each network can then be used as a channel for the protocol being used. Message types that are best suited for a given network are seamlessly sent using the appropriate channel.

The ParaNets architecture has several short-term research challenges and long-term implications which we discuss in this chapter. Some of the short-term challenges include revisiting the transport, routing, addressing, security, and administrative issues in challenged networks. The major long-term implication of ParaNets is that the conventional protocol stack evolves into a more flexible and adaptive *cross-layered protocol tree*. We believe that merging this new concept with the current trend of cross-layer approaches [92], opens the door for protocols capable of providing more robust, timely, and intelligent decisions for challenged networks.

In order to quantitatively show the gain of using ParaNets, we study its impact on Delay Tolerant Mobile Networks (DTMNs) [45]. We use DTMNs as a representative challenged network to evaluate the ParaNets architecture. We validate our short-term challenges and long-term implications by showing the significant improvement in the results of the ParaNets-based solutions when compared to current approaches. Our main goal is to provide a novel architecture over which future challenged networking protocols can be built.

7.2 The ParaNets Architecture

We propose the ParaNets architecture based on the current and future expected trend of individuals accessing multiple networks through one or more devices. We exploit the availability of these devices through their participation in the creation of protocols that can better serve challenged networks. This idea has a profound impact on the services and protocols required to run on such devices and networks.

An instance of the ParaNets architecture is shown in Figure 7.1. The endpoints, A and B, have access to three different networks in parallel. The three

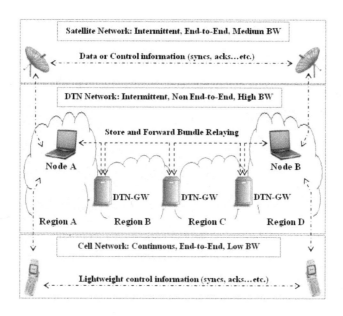

Figure 7.1: A general parallel network architecture.

networks are a classical DTN network [34], a cellular network, and a satellite network. The DTN network represents the challenged network in this case. This network has multiple regions with DTN gateways at the edges of these regions. These gateways perform translation and persistent store-and-forward relaying of "messages bundles" [34]. The DTN network is generally characterized by high bandwidth, but intermittent, and non-end-to-end connectivity. The cellular network, however, is characterized by continuous, end-to-end connectivity, but low bandwidth connections. Finally, the satellite network generally falls in the middle in terms of available bandwidth. It provides intermittent, end-to-end connectivity.

When nodes A and B need to transmit bundle messages over the DTN network, these messages are typically transmitted over the classical DTN network using store-and-forward relaying. Without ParaNets, functionality, such as, routing or reliability, are only provided via the challenged network. Given the characteristics of the challenged network, it becomes increasingly hard to develop robust, efficient solutions. With ParaNets, however, the available parallel cellular and satellite networks are treated as channels that can be used depending on the type of message. For example, the cellular network is best used for transmitting lightweight control information such as acknowledgements, synchronization messages, or routing updates. The satellite network, which usually provides predictable, scheduled, end-to-end connections, can also be used for transmitting control information, or small to average sized message bundles. We finally note that access to the parallel networks is not restricted to the endpoints, but can be extended to the DTN gateways as well, thus enabling extra services for communication in such environments.

7.3 Research Challenges and Implications

There are numerous new research challenges and implications that emerge as a result of ParaNets. Many of these challenges are open for future research. In this section, we identify some of these challenges. We also provide some insight on possible approaches to solve some of these problems. We generally classify these challenges into short-term challenges and long-term implications.

7.3.1 Short-Term Challenges

We now discuss what we believe are the most immediate and crucial research challenges created as a consequence of ParaNets. These challenges include:

Transport: The way in which various transport layer services are currently provided in challenged networks will most likely change with the introduction of ParaNets. Control messages related to these transport services, such as connection establishment, congestion control, and reliability, no longer need to traverse the challenged network. These messages now have an alternative channel over which they can be transmitted. Lightweight data carrying information regarding buffer sizes, acknowledgments, lost packets, and battery power, for example, can now be quickly delivered to far ends of a challenged network. We examine this approach for maintaining end-to-end reliability in our evaluation section.

Routing: Most of the routing approaches adopted in challenged networks rely on opportunistic or scheduled links that are available for short periods of time. Routing decisions are made based on information that can easily be stale due to the nature of the challenged network itself. Information is often disrupted or delayed for long periods of time. ParaNets, in this case, offer much more up-to-date information regarding the challenged network. This information can then lead to more efficient and reliable routing decisions. For example, with ParaNets, information such as node power, path buffers, routes, or location coordinates, can all be sent more quickly and reliably.

Addressing: Since nodes are connected to multiple networks, we need to develop methods by which to correlate and identify the different network addresses given to a node. In other words, we need to ensure that a given node, X, with a given IP, α, is also reachable through the cell number, β, and γ on a third network. New naming schemes and address resolution techniques that can map α to β to γ is yet another interesting research challenge.

Security: Since ParaNets allow operation over multiple networks, there are multiple paths over which security related attacks can be launched. However, these multiple paths also offer alternatives for ensuring authenticity and integrity. Security related information, such as certificates or various keys, can now be transmitted over an appropriate alternate path. Coordinating security measures across such heterogeneous environments is the important challenge.

Administration: Because of ParaNets, a node has access to multiple networks that fall under different administrative systems and domains. Each of these domains has its own operational procedures, administrative policies, and costs.

For example, the traversal of traffic on a network, A, in order to provide services for a network, B, might not be allowed by A. The problem becomes more challenging if A is not designed for such scenarios. For example, the use of the cellular network to send control information might be rejected by cell providers offering voice-only service. The architecture we present takes the current trend of network convergence to new levels with new administrative challenges.

7.3.2 Long-Term Implications

Most protocols are guided and limited by the conventional protocol stack. While the idea of the vertically layered protocol stack has its advantages, this rigid architecture has been broken on many occasions to improve performance. This deviation is particularly prominent in recent research thrusts in cross-layer solutions [92], challenged overlay networks [34], and some multi-network devices [25]. We take these current trends, combine them with ParaNets, and present a vision of how we believe future protocols can be built.

We believe that future protocols will be built based on a tree, shown in Figure 7.2, rather than a fixed stack. The general idea behind the protocol tree is to build protocols in a more flexible and adaptable manner. The layers used in the protocol stack are still the same, but the ways in which they interact are different. Each layer in the tree is built such that the interface between layers are more

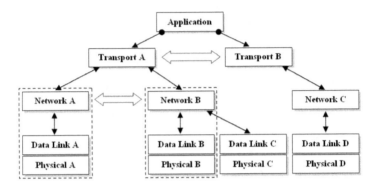

Figure 7.2: The new vision of a protocol tree.

adaptable to a range of possible underlying layers. When a session needs to be established, the most appropriate path, from root to leaf, is chosen, which then represents a temporary virtual stack for the session.

An application is able to use protocols provided by transport layers A or B, as shown in Figure 7.2. In this case, transport protocols A and B are best suited for different underlying networks. Furthermore, each transport protocol has the option of using multiple underlying networks. This is illustrated in Figure 7.2 where transport A can access both networks A and B. We believe that the network, MAC, and physical layers are more tightly coupled in such an architecture. This flexibility in the protocol tree leads us to develop more expandable and backwards compatible protocols, and also provides multiple opportunities for each layer to perform its tasks in the most efficient and effective manner.

The arrows shown in Figure 7.2 represent two concepts. The thin black arrows show how the current trend of cross layering will be maintained in the protocol tree. Sharing information vertically between layers has proven to be very useful in many cases [92], and we believe that this trend will continue, particularly for mobile and challenged networks. On the other hand, the thick horizontal arrows represent what we call cross-network, or cross-stack, cooperation. For example, if a device has connectivity to both networks A and B, routing information gathered from each network can be shared, enabling both layers to learn more than it is capable of knowing on its own. This fact becomes more apparent when we consider ParaNets combined with challenged networks. A node can learn much more about a route to other nodes in the challenged network through the parallel cellular network. We note that the protocol tree vision is more of a long-term implication of both, current research trends, as well as ParaNets. However, more attention needs to be directed towards this implication since we believe that it is a highly probable evolution of the current state of network design.

7.4 Evaluation

The primary goal of our evaluation is to demonstrate the impact ParaNets can have on challenged networks. To fulfill this goal, we use Delay Tolerant

Mobile Networks (DTMNs) as the representative challenged network over which we conduct our evaluation. DTMNs are a special kind of DTNs, where all nodes are mobile, and no end-to-end path necessarily exists between any two nodes [45]. In our evaluation, we incorporate and address some of the short-term challenges mentioned in the previous section, such as reliability and message delivery.

7.4.1 Simulation Environment

We conducted our evaluation using OPNET. We integrated the ParaNets architecture with that of DTMNs by adding the ParaNets handler. This handler coordinates access to multiple networks, which, for our evaluation, were an 802.11-enabled DTMN and a cellular network. We evaluated the performance of the *best* approaches used for message delivery and reliability in a classical DTMN, shown in previous research [45] [44], to that of the ParaNet-enabled DTMN. The best approaches we considered were the *passive*[45] and *active*[44] cures. These cures are approaches used to control and stop message floods in DTMNs by gradually healing message-infected nodes, either through a "passive" kill message or an "active" receipt. The passive and active cures are also used for end-to-end reliability in DTMNs. We particularly show how a ParaNet-enabled DTMN helps spread either one of these cures in a more efficient manner.

In our simulations, we used the random way-point mobility model, but avoided the major problem of node slow down. We used random way-point because our investigation showed that, for this particular case, the mobility model did not affect the relative performance of our solutions. The node speed ranged between 15 to 25 meters per second, and the rest period was 10 seconds. Every point in our results was taken as an average of 20 different seeds. We ran the simulation until the acknowledgment for message delivery was received.

We summarize the major parameters we use in our simulations as follows: *Terrain* is the area over which the *Number of Nodes* are scattered. We run our tests over $5km^2$ with nodes between 25 and 75 (nominal value = 50 nodes). The *Beacon Interval* is the period after which a beacon, used for neighbor discovery, is sent. The beacon interval range is between 10 milliseconds and 10 seconds (nominal value = 0.5 seconds). The *Times-To-Send* (TTS) is the number of times a node will successfully forward a message before it is dropped. This value ranges from 1 to 75 times (nominal value = 50). *Retransmission Wait Time* represents the amount of time a node remains idle after successfully forwarding a message to another node. When the retransmission wait time expires, the node then tries to resend the same message. This time ranges between 0 to 50 seconds (nominal value = 5 seconds). Finally, the *ParaNets Percentage* is the percentage of nodes that have access to a parallel network in order to make use of the ParaNets architecture. This percentage ranges between 2% and 100% (nominal value = 100%).

We use two metrics to evaluate the impact of ParaNets on DTMNs. The first metric is *Delay*, measured by the total time spent to send a bundle message as well as receive its acknowledgement through the cure. We also call this the round trip time of a message. The second metric is *Cost*, measured by the total number of bundle messages introduced into the network.

7.4.2   Simulation Results

We now present a summary of our extended set of simulations, along with a specific subset of our simulation results that clarify and support our goal. All of the results are shown for a single sender sending one bundle message to a single destination. The main goal of these experiments is to study the impact of ParaNets on DTMNs, which helps us better understand its potential impact on challenged networks.

Since both the beacon interval and times-to-send represent the willingness of a node (i.e., how hard a node tries to send a message), we only show results for the beacon interval. The results in Figure 7.3 show the impact of changing the beacon interval. We witness similar performance trends in the three approaches in terms of delay (Figure 7.3(a)) and cost (Figure 7.3(b)). In general, we incur an overall

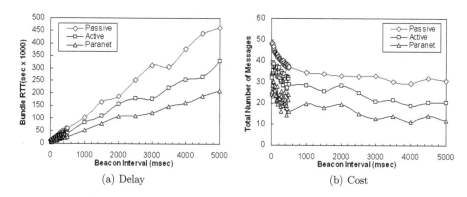

(a) Delay   (b) Cost

Figure 7.3: Impact of beacon interval in ParaNets.

increase in round trip time as the beacon interval increases, and a decrease in the total number of bundles transmitted. The less frequently nodes send beacons, the harder it is for them to discover neighbors to which they can forward a message, resulting in an increase in delay. This loss of connection opportunities, however, is contrasted by an overall decrease in cost since messages are relayed less often.

The second result set is illustrated in Figure 7.4. The figure shows the impact of varying the network density (25, 50, and 75 nodes) on our metrics. For all approaches, as the number of nodes increases, the round trip time decreases and the total number of bundles transmitted increases. As the density increases, more nodes in the network spread the cure faster. This faster spread of the cure, however, comes at the cost of an increase in the total number of messages injected into the network.

In the results in Figures 7.3 and 7.4, we see that the ParaNets-based approach outperforms both the active and passive cure approaches regardless of the change in beacon interval or node density. The reason for this improvement in performance when using ParaNets is that the cellular network acts as a high-speed and low-bandwidth channel. This channel enables the cure to propagate much faster, which leads to two crucial results. First, the round trip time is greatly reduced since the return time is essentially zero because the acknowledgement/cure traverses via the cellular network instead of the challenged network. Second, the total number of messages is much smaller with ParaNets because, once the destination receives the message, the protocol using the ParaNets architecture quickly stops nodes in the network from forwarding the message any further. In other words, the nodes in the network know earlier, via the parallel network, that the message was delivered, and therefore, stop retransmitting the message.

We note that the active approach outperforms the passive approach for both metrics in Figure 7.3. When the cure is "actively" transmitted as a separate message, it reaches the source faster since the uninfected nodes also participate in delivering the active cure. This reduction in return time also stops the nodes in the network from retransmitting earlier than in the passive approach, resulting in a smaller total number of messages transmitted. The less chatty passive

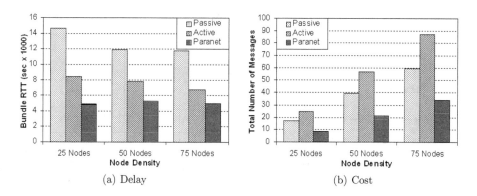

(a) Delay

(b) Cost

Figure 7.4: Impact of number of nodes in ParaNets.

approach, however, is more advantageous than the active one if we include the acknowledgement message as part of the cost, as shown in Figure 7.4(b).

The results we have presented so far assume that all the nodes in the network are ParaNets-enabled. This assumption also explains why we show only one line for ParaNets, because the active and passive approaches running over ParaNets perform the same; whether the cure is active or passive, it reaches all nodes as soon as the message is delivered to the destination. What happens, however, if only some nodes in a challenged network have access to a parallel network?

Figure 7.5 answers this question by showing the impact of having only a subset of nodes ParaNets-enabled. We ensure that the sender node is *not* ParaNets-enabled or else there will be no change in RTT since the sender will receive the cure as soon as the message is delivered to the destination. We now see a difference

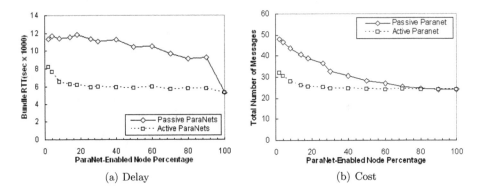

(a) Delay   (b) Cost

Figure 7.5: Impact of ParaNets-enabled node percentage.

in the active vs. passive performance over the ParaNets architecture. The convergence in performance that we see is, again, because as more nodes are ParaNets-enabled, the faster the source receives the cure until no difference exists when all nodes are ParaNets-enabled. Overall, there is a decrease in both delay (Figure 7.5(a)) and cost (Figure 7.5(b)) as the percentage of ParaNets-enabled nodes increases. The interesting observation, however, is that similar improvement in performance is also achieved only with a small number of ParaNets-enabled nodes. This result occurs, as shown, for either the active or passive cure running over the ParaNets architecture.

7.5 Summary

The ParaNets architecture we presented in this chapter aims to provide a basis on which solutions targeted towards challenged networks can be built. Its design exploits the availability of multiple networks in parallel in such a way to help develop more efficient and robust solutions for challenged networks. We have studied the impact of the ParaNets architecture on a representative challenged network, namely, Delay Tolerant Mobile Networks (DTMNs). Solutions based on ParaNets are shown, through our evaluation, to outperform current state of the art solutions, even when only a subset of nodes are ParaNets-enabled.

This work represents the first steps towards developing a large scale, robust, and scalable architecture for challenged networks. This architecture must be flexible and expandable enough to be the basis upon which future solutions for challenged networks can be built. Many other problems and challenges need to be addressed in ParaNets. We have contributed by indicating some short-term research challenges, such as, transport, routing, addressing, security, and administration. We have also demonstrated the evolution of the protocol stack to a cross-layered protocol tree, as a long-term implication of ParaNets. Many challenges, however, remain to be tackled in future work. Nevertheless, we believe that most of the current challenged networks solutions can benefit from the incorporation of ParaNets into their architecture.

Chapter 8

Conclusions

The Internet has revolutionized the way people with which people manage their lives. Its full impact, however, has not yet been realized. User trends and expectations have motivated the evolution of numerous forms of networks. Mobile networks, in particular, have recently witnessed a sharp rise in both demand and usage to satisfy many novel applications. Specifically, more complex applications have emerged that require network connectivity in extreme environments. Examples of these applications include: satellite networks, planetary and interplanetary communications, military/tactical networks, disaster response, disconnected villages, acoustic networks, and other forms of large-scale and sparse mobile networks.

The new networking challenges and assumptions that rise as a result of these novel applications are unprecedented. These challenges include, but are not limited to, network partitioning, intermittent connectivity, large delays, the high cost of infrastructure deployment, and the absence of end-to-end routes. The demand for applications leading to these challenges, along with the evolution in networking technologies, have all contributed to a great change in user expectation. Users expect to be connected in all places and at all times. These novel networking challenges, along with the parallel evolution in networking technologies, has all prompted the recent rise in Delay and Disruption Tolerant Networking (DTN) research.

This dissertation represents a major contribution to DTN research with an overall vision of extending the Internet to the far edges of connectivity, even in the most extreme environments. In this section, we first summarize the work that we present in this dissertation. We then show how this work represents major steps towards fulfilling our vision, objective, and thesis. We also describe the impact of our work on the research community. Finally, we briefly discuss our future work.

8.1   Dissertation Summary

The overall vision we have for this dissertation is to enable the Internet to expand and reach out to communities and environments with great connectivity challenges. This vision helps us set an objective where we contribute to the development of protocols and architectures that enable and facilitate reliable and robust communication in extreme networking environments known as delay and disruption tolerant networks (DTNs). In this dissertation, we pursue this objective by addressing what we believe to be the most pressing challenges in DTNs, namely, routing and data delivery, transport services, and architectural designs. We particularly focus on five major contributions which we present and studied in detail throughout this dissertation. We now present a brief summary for each of these major contributions.

In our first contribution, in Chapter 3, we study the problem of efficient message delivery in delay tolerant sparse mobile networks. We initially propose a new sub area in DTNs known as delay tolerant mobile networks (DTNMs) which focuses on disconnected sparse mobile networks. We also address the problem of message delivery in these environments where minimal assumptions regarding other nodes in the network exist. In such cases, as well as many others, flooding is often used for the purpose of message delivery. The high cost incurred with flooding was the motivating factor for proposing several controlled flooding schemes.

The controlled schemes are integrated in an overlay layer which is built on top of a transport layer in the network. The specific schemes we examine are basic probabilistic, time-to-live, kill time and Passive Cure. We study the impact of these schemes on network efficiency and overall message delivery delay. In our evaluation, via simulations, we demonstrate that for a given sparse mobile network, the schemes reduce the number of messages and beacons sent in the network. This occurs with either no increase or only a small increase in the overall message delay. We then chose the scheme that performs best, which turns out to be a combination of the Passive Cure with time-to-live and basic probabilistic schemes, and stress test it. We examine this combination scheme over an extended architecture that we propose, to accommodate for other real-world scenarios. Specifically, scenarios where stable high-end nodes, such as access points or mesh routers, exist within the network. We show the impact of having even a very small number of such an infrastructure, can greatly reduce the cost of flooding when using our schemes.

After introducing DTMNs, we focus on scenarios where nodes in disconnected mobile networks form clusters which we call regions. In Chapter 4, we particularly study the idea of using a dedicated set of messengers for message delivery in between these disconnected clusters. We introduce two messenger ownership schemes, regional and independent, as well as three scheduling algorithms, periodic, storage-based, and on-demand. These schemes and algorithms combined, provide us with a set of solutions which we then evaluate in details via analysis and simulations.

Our results from studying these schemes and algorithms demonstrate that the choice of a particular ownership and scheduling combination ultimately depends on the environment in which it is deployed. This choice also depends on which metric is most important to the application being served. For example, while the on-demand algorithm promises the least delay, it comes at the expense of large cost and low efficiency. The periodic algorithm, on the other hand, seems to be a reasonable solution as it maintains the most balanced performance with respect to our metrics. Selecting an appropriate periodic interval, though, is a challenge. The storage-based algorithm generally gives the highest efficiency and least cost but requires setting the storage limit, and comes at the expense of high delay. In the end of the chapter, we also show how even the simplest form of adaptive strategies can lead to significant improvement in performance.

We consider transport layer problems in DTNs in Chapter 5. Our main focus in this chapter is on reliability approaches in DTNs. We use DTMNs as a sample DTN on which we study the approaches presented in this chapter. In general, we introduce four different reliability approaches: hop-by-hop, active receipt, passive receipt, and network-bridged receipt. The hop-by-hop approach guarantees only hop-based reliability, but no end-to-end reliability. The other three approaches provide end-to-end reliability at different costs and tradeoffs to the network.

We investigate and evaluate these approaches via simulation. Overall, we discover that the choice of the most suitable reliability approach depends on the expected complexity of the underlying DTN. For example, the hop-by-hop is the simplest approach and consumes the least buffer space, but comes at the tradeoff of no end-to-end reliability. Conversely, the network-bridged receipt provides end-to-end reliability with the least delay, at the expense of the underlying networking complexity. Also, the priority of cost versus delay governs the choice between the active and passive receipt. Overall, the outcome of this study helps us identify which approach can best be used depending on the metric that mostly concerns any given application.

Our attention is then shifted in Chapter 6 to a different form of DTNs that focuses more on disruption and intermittent connectivity rather than large delays. In this chapter we present a Data Bundling System for Intermittent Connections (DBS-IC), a system which deals with intermittent connectivity by proactively delivering data to the mobile user. DBS-IC is comprised of a Stationary Agent (SA), located on a machine with a stable connection to the Internet, and a Mobile Agent (MA), located on an intermittently connected mobile device. By confining the gathering of web, email, and file data to the SA, DBS-IC reduces the data transfer on the mobile device to a bulk TCP data transfer, which allows our system to utilize available bandwidth extremely well.

The overall goal of the system is to enable mobile users to take the best advantage of opportunistic connectivity options that may randomly become available. This goal enables the user to obtain and view required data, even at times when the mobile device is disconnected. We fully implement DBS-IC and find that our system can make data available to the mobile user up to 20 times faster than if the data were traditionally gathered on the mobile device itself.

Since the area of DTNs is still relatively young, we dedicate a large portion of this dissertation to addressing architectural challenges. Even though we had tackled some architectural challenges in DTMNs and DBS-IC, a fresh vision was needed for a generic DTN architecture. We introduce this new vision, named ParaNets, in Chapter 7. The ParaNets architecture aims to provide a basis on which solutions targeted towards challenged networks can be built. Its design exploits the availability of multiple networks in parallel in such a way to help develop more efficient and robust solutions for challenged networks. We study the impact of the ParaNets architecture on a representative challenged network, namely, Delay Tolerant Mobile Networks (DTMNs). Solutions based on ParaNets are shown, through our evaluation, to outperform current state-of-the-art solutions, even when only a subset of nodes are ParaNets-enabled. We believe that ParaNets represents the future trend over which new protocols for challenged networks will be built. Numerous issues such as routing, reliability, security, and network administration, need to be revisited in the light of this new architecture.

8.2 Thesis and Impact

After providing a brief summary for each major contribution in this dissertation, we now show how these contributions collectively fit together to fulfill our vision, objective, and thesis. We had initially started our quest towards developing our vision based on the increased dissemination of mobile devices, the increased need for challenged network applications, and the alternatives available for wireless connectivity. Based on these factors, we envision an expanding Internet that reaches out to remote communities and environments with numerous connectivity challenges, therefore, spurring further research in delay and disruption tolerant networks. As a result of this vision, we set a long term objective to develop and study protocols and architectures that provide us with alternatives to achieve reliable and robust communication in extreme environments.

Even though the long term objective may seem ambitious and hard to reach, we broke it down into a set of manageable challenges that can be addressed by the research community in the short-term. However, with respect to the current state of knowledge and progress in the field, we chose to focus on what we believe to be a crucial subset of these challenges.

The first challenge is to design different architectures over which novel solutions can be built and tested in order to determine the best architectures suitable for any given challenged network. We address this challenge in Chapters 3, 6, and 7. We introduce the novel sub-area of DTMNs, where we consider DTN architectures where nodes are mobile and sparsely scattered. We also present a multi-agent architecture in our DBS-IC system to address frequent disconnections mobile devices might experience. Finally, we propose a novel general architecture for DTNs, known as ParaNets, which takes advantage of multiple networks available to a mobile device.

The second major challenge has to do with routing alternatives and data delivery techniques in the challenged environments where DTNs are assumed to exist. This challenge is tackled in Chapters 3, 4, and 6. We present and study several flooding techniques over DTMNs and examine variations of the DTMN architecture. Furthermore, we consider other scenarios where nodes form clusters, known as regions, that cannot communicate with each other. In these scenarios, we examine using a set of dedicated messengers to establish interregional communication. We extensively study various messenger ownership and scheduling algorithms to identify the algorithms that best suit particular network setups and environments. Finally we address date delivery in other DTN-related scenarios, particularly during frequent short disconnections, through our DBS-IC architecture.

The last important research challenge has to do with the value added services of which the transport services, such as reliability and congestion control, are most critical. This challenge is addressed in Chapters 5 and 6. We focus our attention on reliability approaches and tradeoffs in DTNs. We introduce and examine the tradeoffs between different levels and approaches for reliability in DTNs. We also consider reliability at the application layer in intermittent environments through our DBS-IC system.

At this point we can clearly map the various contributions that we present in this dissertation to the list of challenges we identify based on our goals and vision. We were able to achieve robust message delivery over different DTN architectures via our controlled flooding schemes in DTMNs, messenger scheduling in clustered DTMNs, and DBS-IC in intermittent network settings. Similarly we addressed reliable communication in these environments as well through our study of different reliability approaches for DTNs, as well as application level reliability in DBS-IC. Hence, we believe that after this extensive study, experience and knowledge gained, and results obtained from our work, we can propose a firm thesis for this dissertation. As the thesis of this dissertation, we posit:

The major challenges in delay and disruption tolerant networks, namely, data delivery, transport services, and architectural designs, are crucial problems to which we propose and study alternative solutions to ultimately achieve reliable and robust communication in these networks.

In general, throughout our study for these challenges in DTNs, our work has greatly contributed to the body of research in challenged and extreme networks. Our work, presented in this dissertation has helped advance the state of the art work in delay and disruption tolerant networks. We have contributed in pinpointing and solving crucial problems towards enabling network communication in challenged environments. The impact of our work can be summarized in the following points:

- With the introduction of DTMNs and addressing controlled flooding issues in these environments, we introduced a new sub-area within the DTN researchers that has continued to evolve in the past couple of years. As evidence of this evolution, we note that a conference has been created to address this area.

- We are one of the first groups to make a focus shift in the work presented in DTN community by examining transport layer issues and comparing different reliability approaches in DTN environments. Our work and interaction with other researchers in the area has started a new thrust away from routing and message delivery in DTNs.

- We propose ParaNets, a novel general architecture for DTNs, and expect it to be a solid base over which many future solutions and protocols for DTNs will be developed. Based on the feedback received from the research community, ParaNets will be a major focus in the future.

- We have influenced the direction and evolution of current and future research in DTNs by getting feedback and communicating our work to the area leaders in DTNs. Through our interaction with these top researchers in IETF meetings, conferences, and workshops, we were able to share and impact the evolution of the work in this area.

- To date, our work has been referenced in many papers, theses, and dissertations. Furthermore, some of our papers are taught in seminars or courses that address DTNs.

8.3   Future Work

The area of challenged networks and DTNs remains relatively young, and has only recently gained great momentum. Hence, research on challenged networks has only begun, and we believe it will continue to evolve as an important area of research in computer networks. Through challenged networks, we can push the Internet into the reaches of the planet not currently connected.

Having spawned several research threads in this dissertation, there is still work that remains to be conducted and numerous unanswered questions regarding challenged networks. For example, with respect to DTMNs, future work in this thread includes addressing security issues, since such networks are subject to denial of service attacks. Also, since flooding-based schemes do not perform well in dense environments, measures to help nodes modify their behavior when they enter densely populated areas need to be developed.

Considering messenger scheduling in clustered mobile networks, future work in this area includes studying environments where messengers are destroyed or lost. Hence value added services, such as reliability and fault tolerance would need to be tackled in these cases. Also, the algorithms we propose represent the first generation of scheduling strategies. More intelligent and adaptive solutions can be designed where messengers can dynamically chose a different scheduling algorithm, or even a mix of strategies, to enable them to adapt to changing network conditions. In the end, as networks are increasingly expected to provide communication in hostile and challenged environments, more robust solutions will be needed to fulfill such expectations.

We consider our work on reliability approaches an initial step in the investigation of transport layer issues in DTNs. Future work in this area is to apply these approaches to different forms of DTNs, and examine how they might be modified and applied to other DTN architectures. Also, limited buffers or storage space is a new rising concern that presses for efforts addressing other transport layer issues, particularly, congestion control. In fact, queuing and buffer management, as well as message bundle priorities, are other services that require further research.

With respect to intermittent connectivity, we believe that DBS-IC is a solid step to improve mobile users' experience in the face of intermittent connectivity. However, there are several areas for future work. A valuable improvement to our system would be the addition of an intelligent gathering and bundling agent. This agent would be located on the SA and would use past viewing trends to dynamically decide which data the user might need in the future. Our system could further be extended to handle interactive data, caching user requests during times of disconnection. This extension would be especially useful with interactive web pages that require user input, and with newly composed emails that the user wishes to send. Built on top of our work, these improvements would help make the in-motion mobile user's experience almost equal to that of a stationary user's.

We envision our work on ParaNets as a major point after which the philosophy of future solutions for challenged networks may change. This work represents the first steps towards developing a robust, scalable, and most importantly, adaptive, architecture for challenged networks. This architecture must be flexible enough to be the basis upon which future solutions for challenged networks can be built. Many other problems and challenges need to be addressed in ParaNets. We have contributed by indicating some short-term research challenges, such as, transport, routing, addressing, security, and administration. We have also demonstrated the evolution of the protocol stack to a cross-layered protocol tree, as a long-term implication of ParaNets. Many challenges, however, remain to be tackled in future work.

Overall, the ParaNets architecture is a novel direction that opens the door for much additional research, both in challenged networks, and in conventional networks. The spread of devices that can take advantage of multiple networks, to which they have access, is very appealing, and is a new direction many researchers are currently pursuing. Recent protocols and architectures have focused mostly on operating over a singe network. The ParaNets architecture, however, changes this notion, and leads to a great opportunity to not only rethink current solutions to make them adapt to multiple networks, but also to develop new solutions tailored to the new trends.

In the end, observing the big picture, the various thrusts within computer networking research are starting to converge. With the current trends we are witnessing, this convergence is characterized by an overlap and dependency that is becoming much more prominent than before. For example, it is currently difficult to address problems in ad hoc or mesh networks without considering distributed cooperation and information dissemination between nodes, bandwidth allocation, congestion control, channel scheduling and allocation, disconnections and network partitioning, single hop and end-to-end reliability, protocol vulnerability and security, and the impact of all of these factors when attached to a wired network infrastructure. This interdependency between the different sub areas in computer networking research, we believe, will govern the philosophy of future research not only in DTNs, but in computer networking in general.

Bibliography

[1] University of South Florida: Center for robot-assisted search and rescue. http://crasar.csee.usf.edu/.

[2] Technology and Infrastructure for Emerging Regions. http://tier.cs.berkeley.edu/.

[3] The Wizzy project. http://www.wizzy.org.za/.

[4] First Mile Solutions. http://www.firstmilesolutions.com/.

[5] DTNRG. Delay Tolerant Networking Research Group. http://www.dtnrg.org/.

[6] IPNRG. Inter-Planetary Networking Spcial Interest Group. http://www.ipnsig.org/.

[7] The Drive-thru Internet project. http://www.drive-thru-internet.org/.

[8] The DHARMA project. http://dharma.cis.upenn.edu/.

[9] The Archive for the ACM Workshop on Challenged Networks (CHANTS). http://chants.cs.ucsb.edu/.

[10] D. G. Andersen, H. Balakrishnan, and F. Kaashoek. Improving Web Availability for Clients with MONET. In *2nd Symp. on Networked Systems Design and Implementation (NSDI)*, Boston, MA, May 2005.

[11] A. Balasubramanian, B. N. Levine, and A. Venkataramani. DTN Routing as a Resource Allocation Problem. In *ACM SIGCOMM*, Kyoto, Japan, August 2007.

[12] B. Bellur and R. Ogier. A Reliable, Efficient Topology Broadcast Protocol for Dynamic Networks. In *IEEE INFOCOM*, New York, NY, March 1999.

[13] C. Bettstetter, H. Hartenstein, and X. Perez-Costa. Stochastic properties of the random waypoint mobility model. *Wireless Networks*, 10(5):555–567, 2004.

[14] G. W. Boehlert, D. P. Costa, D. E. Crocker, P. Green, T. OBrien, S. Levitus, and B. J. L. Boeuf. Autonomous Pinniped Environmental Samplers; Using Instrumented Animals as Oceanographic Data Collectors. *Journal of Atmospheric and Oceanic Technology*, 18(11):1882–1893, 2001.

[15] E. Brewer, M. Demmer, B. Du, M. Ho, M. Kam, S. Nedevschi, J. Pal, R. Patra, S. Surana, and K. Fall. The Case for Technology in Developing Regions. *Computer*, 38(6):25–38, 2005.

[16] J. Burgess, G. Bissias, M. D. Corner, and B. N. Levine. Surviving Attacks on Disruption-Tolerant Networks without Authentication. In *ACM MobiHoc*, Montreal, Canada, September 2007.

[17] J. Burgess, B. Gallagher, D. Jensen, and B. N. Levine. MaxProp: Routing for Vehicle-Based Disruption-Tolerant Networking. In *IEEE INFOCOM*, Barcelona, Spain, April 2006.

[18] S. Burleigh, A. Hooke, L. Torgerson, K. fall, V. Cerf, B. Durst, K. Scott, and H. Weiss. Delay-Tolerant Networking: An Approach to Interplanetary Internet. *IEEE Communications*, 41(6):128–136, June 2003.

[19] S. Burleigh, E. Jennings, and J. Schoolcraft. Autonomous Congestion Control in Delay-Tolerant Networks. In *International Conference on Space Operations*, Rome, Italy, June 2006.

[20] B. Burns, O. Brock, and B. Levine. MV Routing and Capacity Building in Disruption Tolerant Networks. In *IEEE INFOCOM*, Miami, FL, March 2005.

[21] B. Burns, O. Brock, and B. N. Levine. Autonomous Enhancement of Disruption Tolerant Networks. In *IEEE International Conference on Robotics and Automation (ICRA)*, Orlando, FL, May 2006.

[22] B. Burns, O. Brock, and B. N. Levine. MORA Routing and Capacity Building in Disruption-Tolerant Networks. *Elsevier Ad hoc Networks Journal*, To Appear, 2008.

[23] V. Cerf and R. Kahn. A Protocol for Packet Network Intercommunication. *IEEE Transactions on Communications*, May 1974.

[24] V. Cerf, et al. Interplanetary Internet (IPN): Architectural Definition. *IETF Internet Draft, draft-irtf-ipnrg-arch-00.txt*, May 2001.

[25] R. Chandra, P. Bahl, and P. Bahl. MultiNet: Connecting to Multiple IEEE 802.11 Networks Using a Single Wireless Card. In *IEEE INFOCOM*, Hong Kong, March 2004.

[26] H. Chang, C. Tait, N. Cohen, M. Shapiro, S. Mastriann, R. Floyd, B. Housel, and D. Lindquist. Web Browsing in a Wireless Environment: Disconnected and Asynchronous Operation in ARTour Web Express. In *ACM/IEEE International conference on mobile computing and networking*, Budapest, Hungary, 1997.

[27] L.-J. Chen, C.-H. Yu, T. Sun, Y.-C. Chen, and H. hua Chu. A Hybrid Routing Approach for Opportunistic Networks. In *ACM SIGCOMM workshop on Challenged Networks (CHANTS)*, Pisa, Italy, September 2006.

[28] M. Chuah, L. Cheng, and B. Davison. Enhanced Disruption and Fault Tolerant Network Architecture for Bundle Delivery (EDIFY). In *IEEE Globecom*, St. Louis, MO, November 2005.

[29] D. Clark. The Design Philosophy of The DARPA Internet Protocols. *Computer Communication Review*, 18(4):106–114, September 1988.

[30] D. Clark and D. Tennenhouse. Architectural Considerations for A New Generation of Protocols. *Computer Communication Review*, 20(4):200–208, September 1990.

[31] T. Clausen and P. Jacquet. Optimized Link State Routing Protocol (OLSR). In *RFC 3626, IETF Network Working Group*, October 2003.

[32] J. Davis, A. Fagg, and B. Levine. Wearable Computers as Packet Transport Mechanisms in Highly-Partitioned Ad-Hoc Networks. *In International Symposium on Wearable Computing*, October 2001.

[33] S. Doshi and T. Brown. Minimum Energy Routing Schemes for a Wireless Ad Hoc Network. In *IEEE INFOCOM*, New York, NY, June 2002.

[34] K. Fall. A Delay-Tolerant Network Architecture for Challenged Internets. In *ACM SIGCOMM*, Karlsruhe, Germany, August 2003.

[35] K. Fall, W. Hong, and S. Madden. Custody Transfer for Reliable Delivery in Delay Tolerant Networks. *Intel Research, Berkeley-TR-03-030*, July 2003.

[36] S. Farrell, V. Cahill, D. Geraghty, I. Humphreys, and P. McDonald. When TCP Breaks: Delay- and Disruption- Tolerant Networking. *IEEE Internet Computing*, 10(4):72–78, 2006.

[37] R. Gass, J. Scott, and C. Diot. Measurements of In-Motion 802.11 Networking. In *Workshop on Mobile Computing Systems and Applications (WMCSA)*, Semiahmoo Resort, WA, April 2006.

[38] J. Gomez, A. Campbell, M. Naghshineh, and C. Bisdikian. Power-Aware Routing in Wireless Packet Networks. In *IEEE International Workshop on Mobile Multimedia Communications (MOMUC99)*, San Diego, CA, November 1999.

[39] M. Grossglauser and D. Tse. Mobility Increases the Capacity of Ad-hoc Wireless Networks. In *IEEE INFOCOM*, Anchorage, AK, April 2001.

[40] M. Grossglauser and M. Vetterli. Locating Nodes with EASE: Last Encounter Routing for ad hoc networks through mobility diffusion. In *IEEE INFOCOM*, San Francisco, CA, March 2003.

[41] M. Grossglauser and M. Vetterli. Locating Mobile Nodes with EASE: Learning Efficient Routes from Encounter Histories Alone. *IEEE/ACM Transactions on Networking*, 14(3):457–469, 2006.

[42] Z. Haas and M. Pearlman. The Performance of Query Control Schemes for Zone Routing Protocol. In *ACM SIGCOMM*, pages 167–177, Vancouver, Canada, August 1998.

[43] K. Harras and K. Almeroth. Inter-Regional Messenger Scheduling in Delay Tolerant Mobile Networks. In *IEEE World of Wireless, Mobile and Multimedia Networks (WoWMoM)*, Buffalo, NY, June 2006.

[44] K. Harras and K. Almeroth. Transport Layer Issues in Delay Tolerant Mobile Networks. In *IFIP Networking*, Coimbra, Portugal, May 2006.

[45] K. Harras, K. Almeroth, and E. Belding-Royer. Delay Tolerant Mobile Networks (DTMNs): Controlled Flooding Schemes in Sparse Mobile Networks. In *IFIP Networking*, Waterloo, Canada, May 2005.

[46] K. Harras, M. Wittie, K. Almeroth, and E. Belding. ParaNets: A Parallel Network Architecture for Challenged Networks. In *IEEE HotMobile*, Tucson, AZ, February 2007.

[47] H. Hassanein and A. Zhou. Routing With Load Balancing in Wireless Ad Noc Networks. In *ACM Workshop on Modeling, Analysis, and simulation of Wireless and Mobile Systems*, Rome, Italy, July 2001.

[48] C. Holman, K. Harras, K. Almeroth, and E. Belding. A Proactive Data Bundling System for Intermittent Mobile Connections. In *IEEE SECON*, Reston, VA, September 2006.

[49] A. Hooke. The Interplanetary Internet. *Communications of the ACM*, 44(9):38–40, September 2001.

[50] P. Jacquet, P. Muhlethaler, T. Clausen, A. Laouiti, A. Qayyum, and L. Viennot. Optimized Link State Routing Protocol for Ad Hoc Networks. In *IEEE Multi Topic Conference (INMIC)*, Lahore, Pakistan, December 2001.

[51] S. Jain, M. Demmer, R. Patra, and K. Fall. Using Redundancy to Cope with Failures in a Delay Tolerant Network. In *ACM SIGCOMM*, Philadelphia, PA, August 2005.

[52] S. Jain, K. Fall, and R. Patra. Routing in a Delay Tolerant Network. In *ACM SIGCOMM*, Portland, OR, August 2004.

[53] D. Johnson and D. Maltz. *Dynamic Source Routing in Ad Hoc Wireless Networks*, volume 353. Kluwer Academic Publishers, 1996.

[54] E. Jones, L. Li, and P. Ward. Practical Routing in Delay-Tolerant Networks. In *ACM SIGCOMM Workshop on Delay-tolerant Networking (WDTN)*, Philadelphia, PA, August 2005.

[55] P. Juang, et al. Energy-Efficient Computing for Wildlife Tracking: Design Tradeoffs and Early Experiences With ZebraNet. In *In International Conference on Architectural Support for Programming Languages and Operating Systems*, San Jose, CA, October 2002.

[56] H. Jun, M. H. Ammar, M. D. Corner, and E. W. Zegura. Hierarchical Power Management in Disruption Tolerant Networks with Traffic-Aware Optimization. In *ACM SIGCOMM workshop on Challenged Networks (CHANTS)*, Pisa, Italy, September 2006.

[57] M. Kaddour and L. Pautet. A Middleware for Supporting Disconnections and Multi-network Access in Mobile Environments. In *PERCOM*, Orlando, FL, March 2004.

[58] G. Karlsson, V. Lenders, and M. May. Delay-Tolerant Broadcast. In *ACM SIGCOMM workshop on Challenged Networks (CHANTS)*, Pisa, Italy, September 2006.

[59] Y.-B. Ko and N. Vaidya. Location-Aided Routing (LAR) in Mobile Ad hoc Networks. In *ACM MobiCom*, pages 66–75, Dallas, TX, October 1998.

[60] Y.-B. Ko and N. H. Vaidya. Location-Aided Routing in Mobile Ad Hoc Networks. *ACM Wireless Networks Journal*, 6(4):307–321, 2000.

[61] E. Krotkov and J. Blitch. The Defense Advanced Research Projects Agency (DARPA) Tactical Mobile Robotics Program. *The International Journal of Robotics Research*, 18(7):769–776, July 1999.

[62] P. Kulkarni, P. Shenoy, and K. Ramamritham. Handling Client Mobility and Intermittent Connectivity in Mobile Web Accesses. In *MDM 2003*, pages 401–407, Melbourne, Australia, January 2003.

[63] K. Lakshminarayanan, I. Stoica, and K. Wehrle. Support for Service Composition in i3. In *ACM Multimedia*, New York, NY, October 2004.

[64] S.-J. Lee and M. Gerla. Dynamic Load-Aware Routing in Ad Hoc Networks. In *IEEE International Conference on Communications (ICC 2001)*, Helsinki, Finland, June 2001.

[65] J. Leguay, T. Friedman, and V. Conan. DTN Routing in a Mobility Pattern Space. In *ACM SIGCOMM Workshop on Delay-tolerant Networking (WDTN)*, Philadelphia, PA, August 2005.

[66] J. Leguay, T. Friedman, and V. Conan. Evaluating Mobility Pattern Space Routing. In *IEEE INFOCOM*, Barcelona, Spain, April 2006.

[67] J. Leguay, A. Lindgren, J. Scott, T. Friedman, and J. Crowcroft. Opportunistic Content Distribution in an Urban Setting. In *ACM SIGCOMM workshop on Challenged Networks (CHANTS)*, Pisa, Italy, September 2006.

[68] Q. Li and D. Rus. Sending Messages to Mobile Users in Disconnected Ad-Hoc Wireless Networks. In *ACM MobiCom*, pages 44–55, Boston, MA, August 2000.

[69] Y. Liao, K. Tan, Z. Zhang, and L. Gao. Modeling Redundancy-based Routing in Delay Tolerant Networks. In *IEEE Consumer Communications and Networking Conference*, Las Vegas, NV, January 2007.

[70] A. Lindgren, A. Doria, and O. Scheln. Probabilistic Routing in Intermittently Connected Networks. In *ACM MobiHoc*, Annapolis, MD, June 2003.

[71] Y. Mao, B. Knuttson, H. Lu, and J. M. Smith. Dharma: Distributed home agent for robust mobile access. In *IEEE INFOCOM*, Miami, FL, March 2005.

[72] A. M.Keller, O. Densmore, W. Huang, and B. Razavi. Zippering: Managing Intermittent Connectivity in DIANA. *Mobile Networks and Applications*, 2(4):357–364, December 1997.

[73] M. Musolesi, S. Hailes, and C. Mascolo. Adaptive Routing for Intermittently Connected Mobile Ad Hoc Networks. In *IEEE World of Wireless, Mobile and Multimedia Networks (WoWMoM)*, Messina, Italy, June 2005.

[74] D. Nain, N. Petigara, and H. Balakrishnan. Integrated Routing and Storage for Messaging Applications in Mobile Ad Hoc Networks. In *Modeling and Optimization in Mobile, Ad Hoc and Wireless Networks (WiOpt)*, Sophia-Antipolis, France, March 2003.

[75] J. Ott and D. Kutscher. Drive-thru Internet: IEEE 802.11b for Automobile Users. In *IEEE INFOCOM*, Hong Kong, March 2004.

[76] J. Ott and D. Kutscher. A Disconnection-Tolerant Transport for Drive-thru Internet Environments. In *IEEE INFOCOM*, Miami, FL, March 2005.

[77] J. Ott and D. Kutscher. A Mobile Access Gateway for Managing Intermittent Connectivity. In *IST Mobile and Wireless Communication Summit*, Dresden, Germany, June 2005.

[78] J. Ott and D. Kutscher. A Modular Access Gateway for Managing Intermittent Connectivity in Vehicular Communications. *European Transactions on Telecommunications*, 17(2):159–174, 2006.

[79] J. Ott, D. Kutscher, and C. Dwertmann. Integrating DTN and MANET Routing. In *ACM SIGCOMM workshop on Challenged Networks (CHANTS)*, Pisa, Italy, September 2006.

[80] V. Park and M. Corson. A Highly Adaptive Distributed Routing Algorithm for Mobile Wireless Networks. In *IEEE INFOCOM*, Kobe, Japan, April 1997.

[81] C. Perkins. Ad-hoc On-Demand Distance Vector Routing. In *IEEE Workshop on Mobile Computing Systems and Applications*, pages 90–100, New Orleans, LA, February 1999.

[82] C. Perkins, E. Belding-Royer, and S. Das. Ad hoc On-Demand Distance Vector (AODV) Routing. In *RFC 3561, IETF Network Working Group*, July 2003.

[83] C. Perkins and P. Bhagwat. Highly Dynamic Destination-Sequenced Distance-Vector Routing (DSDV) for Mobile Computers. In *ACM SIGCOMM*, pages 234–244, London, England, October 1994.

[84] E. Royer and C. Toh. A Review of Current Routing Protocols for Ad-hoc Mobile Wireless Networks. *IEEE Personal Communications Magazine*, 6(2):46–55, April 1999.

[85] J. Scott, P. Hui, J. Crowcroft, and C. Diot. Haggle: A Networking Architecture Designed Around Mobile Users. In *IFIP Conference on Wireless On demand Network Systems and Services (WONS 2006)*, Les Menuires, France, January 2006.

[86] M. Seligman, K. Fall, and P. Mundur. Alternative Custodians for Congestion Control in Delay Tolerant Networks. In *ACM SIGCOMM workshop on Challenged Networks (CHANTS)*, Pisa, Italy, September 2006.

[87] A. Seth and S. Keshav. Practical Security for Disconnected Nodes. In *Workshop on Secure Network Protocols (NPSec)*, Boston, MA, November 2005.

[88] R. Shah, S. Roy, S. Jain, and W. Brunette. Data MULEs: Modeling a Three-Tier Architecture for Sparse Sensor Networks. In *In IEEE International Workshop on Sensor Network Protocols and Applications*, Anchorage, AK, 2003.

[89] T. Small and Z. Haas. The Shared Wireless Infostation Model - A New Ad Hoc Networking Paradigm (or Where There is a Whale, There is a Way). In *ACM MobiHoc*, Annapolis, MD, June 2003.

[90] T. Small and Z. Haas. Resource and Performance Tradeoffs in Delay-Tolerant Wireless Networks. In *ACM SIGCOMM Workshop on Delay-tolerant Networking (WDTN)*, Philadelphia, PA, August 2005.

[91] T. Spyropoulos, K. Psounis, and C. Raghavendra. Spray and Wait: An Efficient Routing Scheme for Intermittently Connected Mobile Networks. In *ACM SIGCOMM Workshop on Delay-tolerant Networking (WDTN)*, Philadelphia, PA, August 2005.

[92] V. Srivastava and M. Motani. Cross-layer Design: A Survey and the Road Ahead. *IEEE Communications*, 43(12):112–119, 2005.

[93] I. Stoica, D. Adkins, S. Zhuang, S. Shenker, and S. Surana. Internet Indirection Infrastructure. In *ACM SIGCOMM*, Pittsburgh, PA, August 2002.

[94] J. Su, A. Chin, A. Popivanova, A. Goel, and E. de Lara. User Mobility for Opportunistic Ad-Hoc Networking. In *IEEE Workshop on Mobile Computing Systems and Applications (WMCSA)*, English Lake District, UK, December 2004.

[95] K. Tan, Q. Zhang, and W. Zhu. Shortest Path Routing in Partially Connected Ad Hoc Networks. In *IEEE Globecom*, San Francisco, CA, December 2003.

[96] C.-K. Toh. Associativity Based Routing for Ad-Hoc Mobile Networks. *Wireless Personal Communications Journal, Special Issue on Mobile Networking and Computing Systems*, 4(2):103–139, March 1997.

[97] A. Vahdat and D. Becker. Epidemic Routing for Partially Connected Ad Hoc Networks. *Technical Report CS-200006, Duke University*, April 2000.

[98] H. Waisanen, D. Shah, and M. Dahleh. Optimal Delay in Networks with Controlled Mobility. In *17th International Symposium on Mathematical Theory of Networks and Systems*, Kyoto, Japan, July 2006.

[99] W. Wang, V. Srinivasan, and K.-C. Chua. Using Mobile Relays to Prolong the Lifetime of Wireless Sensor Networks. In *ACM MobiCom*, Cologne, Germany, August 2005.

[100] Y. Wang, S. Jain, M. Martonosi, and K. Fall. Erasure-Coding Based Routing for Opportunistic Networks. In *ACM SIGCOMM Workshop on Delay-tolerant Networking (WDTN)*, Philadelphia, PA, August 2005.

[101] Y. Wang, C.-Y. Wan, M. Martonosi, and L.-S. Peh. Transport Layer Approaches for Improving Idle Energy in Challenged Sensor Networks. In *ACM SIGCOMM workshop on Challenged Networks (CHANTS)*, Pisa, Italy, September 2006.

[102] Y. Wang and H. Wu. DFT-MSN: The Delay/Fault-Tolerant Mobile Sensor Network for Pervasive Information Gathering. In *IEEE INFOCOM*, Barcelona, Spain, April 2006.

[103] J. Widmer and J.-Y. L. Boudec. Network Coding for Efficient Communication in Extreme Networks. In *ACM SIGCOMM Workshop on Delay-tolerant Networking (WDTN)*, Philadelphia, PA, August 2005.

[104] X. Zeng, R. Bagrodia, and M. Gerla. Glomosim: A Library for Parallel Simulation of Large-Scale Wireless Networks. In *ACM PADS*, Banff, Canada, May 1998.

[105] X. Zhang, J. Kurose, B. N. Levine, D. Towsley, and H. Zhang. Study of a Bus-Based Disruption Tolerant Network: Mobility Modeling and Impact on Routing. In *ACM Mobicom*, Montreal, Canada, September 2007.

[106] X. Zhang, G. Neglia, J. Kurose, and D. Towsley. Performance Modeling of Epidemic Routing. In *IFIP Networking*, Coimbra, Portugal, May 2006.

[107] Z. Zhang. Routing in Intermittently Connected Mobile Ad Hoc Networks and Delay Tolerant Networks: Overview and Challenges. *IEEE Communications Surveys and Tutorials*, 8(1):24–37, 2006.

[108] W. Zhao and M. Ammar. Proactive Routing in Highly-Partitioned Wireless Ad Hoc Networks. In *The IEEE International Workshop on Future Trends of Distributed Computing Systems*, San Juan, Puerto Rico, May 2003.

[109] W. Zhao, M. Ammar, and E. Zegura. A Message Ferrying Approach for Data Delivery in Sparse Mobile Ad Hoc Networks. In *ACM MobiHoc*, Tokyo, Japan, May 2004.

[110] W. Zhao, M. Ammar, and E. Zegura. Controlling the Mobility of Multiple Data Transport Ferries in a Delay-Tolerant Network. In *IEEE INFOCOM*, Miami, FL, March 2005.

[111] W. Zhao, M. Ammar, and E. Zegura. Multicasting in Delay Tolerant Networks: Semantic Models and Routing Algorithms. In *ACM SIGCOMM Workshop on Delay-tolerant Networking (WDTN)*, Philadelphia, PA, August 2005.

[112] S. Zhuang, K. Lai, I. Stoica, R. Katz, and S. Shenker. Host Mobility using an Internet Indirection Infrastructure. In *ACM/USENIX Mobisys*, San Francisco, CA, May 2003.

VDM
Verlag
Dr. Müller

Wissenschaftlicher Buchverlag bietet

kostenfreie

Publikation

von

wissenschaftlichen Arbeiten

Diplomarbeiten, Magisterarbeiten, Master und Bachelor Theses
sowie Dissertationen, Habilitationen und wissenschaftliche Monographien

Sie verfügen über eine wissenschaftliche Abschlußarbeit zu aktuellen oder zeitlosen
Fragestellungen, die hohen inhaltlichen und formalen Ansprüchen genügt,
und haben **Interesse an einer honorarvergüteten Publikation**?

Dann senden Sie bitte erste Informationen über Ihre Arbeit per Email
an info@vdm-verlag.de. Unser Außenlektorat meldet sich umgehend bei Ihnen.

VDM Verlag Dr. Müller Aktiengesellschaft & Co. KG
Dudweiler Landstraße 125a
D - 66123 Saarbrücken

www.vdm-verlag.de

www.ingramcontent.com/pod-product-compliance
Lightning Source LLC
LaVergne TN
LVHW022308060326
832902LV00020B/3347